THE
HUMAN BODY —
Accident or Design?

EXPANDED EDITION

Wayne Jackson

Courier Publications
Stockton, California

THE HUMAN BODY —
Accident or Design?

©1993, 2000 by Courier Publications

ISBN 0-9678044-3-4

Additional copies may be ordered from:

Courier Publications
P.O. Box 55265
Stockton, CA 95205
http://www.christiancourier.com

DEDICATION

This book is affectionately dedicated
to my many friends
who have encouraged and supported me
as I have attempted, through my writings,
to glorify my MAKER.

"For God is not unrighteous to forget your work and
the love which you showed toward his name, in that
you ministered unto the saints, and still do minister."

HEBREWS 6:10

TABLE OF CONTENTS

PREFACE   *i*

FORWARD   *iii*

CHAPTER 1   *1*
THE LAW OF TELEOLOGY   1
THE NATURE OF DESIGN   1
BIOLOGICAL LIFE   2
BIBLICAL VIEW OF THE HUMAN BODY   4
ORGANIZATION OF THE BODY   5
THE CELL   6
D.N.A.   9

CHAPTER 2   *17*
THE SKIN SYSTEM   18
THE SKELETAL SYSTEM   21
THE MUSCULAR SYSTEM   25

CHAPTER 3   *35*
THE DIGESTIVE SYSTEM   36
THE CIRCULATORY SYSTEM   42

CHAPTER 4   *51*
THE NERVOUS SYSTEM   52
THE BRAIN   54
SENSORY ORGANS — THE EYE   59
SENSORY ORGANS — THE EAR   62

CHAPTER 5   *67*
THE RESPIRATORY SYSTEM   67
THE EXCRETORY SYSTEM   72
THE ENDOCRINE SYSTEM   77

CHAPTER 6   83

THE REPRODUCTIVE COMPONENTS   85
CONCEPTION   88
GESTATION   95
THE BIRTH PROCESS   97
THE NEWBORN   99

CHAPTER 7   103

ARE THERE "DESIGN" FLAWS?   103
MAN'S IGNORANCE   103
DESIGN UNRECOGNIZED   104
DEGENERATION   105
ATHEISM'S "WISDOM"   106

APPENDIX I: SLEEP — AN EVIDENCE OF DIVINE DESIGN   111

SCIENCE BAFFLED   111
THEORIES REGARDING SLEEP   112
DIVINE DESIGN   114
BENEFITS OF SLEEP   114
WHERE THE EVIDENCE POINTS   117

APPENDIX II: THE "EYE" OF THE EVOLUTIONARY STORM   119

DAWKINS' CHARGE   120
DENTON'S NEW STUDY   120

APPENDIX III: THE GOD WHO HEALS   123

A PROMISE OF HEALING   123
MIRACULOUS HEALING   124
HEALING BY DESIGN   126
FACILITATING GOD'S HEALING PROCESS 129

APPENDIX IV: SOME QUICK STATISTICS ABOUT THE HUMAN BODY   131

PREFACE

A thousand years before the birth of Jesus Christ, the Lord's illustrious ancestor, David, the king of Israel, declared: "I am fearfully and wonderfully made" (Psalm 139:14). The Hebrew psalmist could not have known, by natural research, how accurate his statement was. The key word is "made." And that implies a Maker.

The research for this book consumed many months. Progressively the author has been profoundly amazed at the countless examples of "design" to be discovered within the human body. There is simply no way it could have evolved by natural processes. The human body is a superlative argument for creation!

I am indebted to several people for their assistance in this production. My son, Jared Hardeman Jackson, is invaluable for his skill in computer graphics. Dr. Bert Thompson of Montgomery, Alabama, proof-read the initial manuscript and made a number of recommendations, as did my esteemed friend, Dr. James Clark (Ob/Gyn – retired) of Boyd, Texas. I am likewise grateful to Dr. Harrell Dodson, a respected surgeon/professor (retired) of Oklahoma City, who read this work, offered several helpful suggestions, and who penned the Foreword.

I am deeply thankful for the many expressions of gratitude since the first edition was published in 1993. I send forth this updated version with the fond hope that increasing good, to God's glory, will be accomplished.

Wayne Jackson
December, 2000

FORWARD

THE HUMAN BODY - Accident or Design? is undoubtedly, the best book on the subject that I have read. The anatomic and physiologic facts recorded are extremely accurate and minutely detailed, indicative if the extensive research involved in accumulating this material.

Furthermore, these scientific facts are presented in language that can be understood by all readers. The repeated emphasis upon the intricate and precise design of the human body as proof of the existence of a designer, keeps the reader focused upon the message of the book.

Also the descriptions of the complexities and inter-relationships of the various cells, organs and systems of the body should make it evident that only God could be its designer. The author even alludes to specific functions of the human body that would be impossible to develop by evolution from another form of life, and to unique features of the design of man's body to equip him for the purposes ordained by God.

Although these physical characteristics of the human body confirm its origin by Divine creation, the most striking confirmation involves the realm in which man is created in God's own image, namely, man's intellect, his will and his soul-endowments that God reserved for man only. Unlike physical characteristics, these spiritual characteristics cannot be weighed, measured, or even located in the human body.

Although God has allowed man, through scientific

endeavors, to develop crude and grossly inferior substitutes for parts of his physical body, the spiritual nature of man, like life itself, will forever remain the realm of God.

Harrell C. Dodson, Jr., M.D., F.A.C.S.
Clinical Professor of Surgery (Retired)
University of Oklahoma College of Medicine

CHAPTER 1

There is a principle in logic that may be called the "Law of Teleology." Teleology has to do with design. The law, simply stated, is this: *Where there is design, there must be a designer.* Even unbelievers have been forced to acknowledge this principle. Paul Ricci, a skeptical professor of philosophy and logic, has written: "'Everything designed has a designer' is an analytically true statement" (1986, p 190).

Since design demands a designer, it necessarily follows that if design is discovered in the fabric of Earth's environment, one would have to conclude, if intellectually honest, that there must be a grand Designer ultimately responsible for this circumstance.

But how is "design" to be defined? Design, at least in part, has to do with the *arrangement of individual components within an object so as to accomplish a functional or artistic purpose.* An automobile contains design because its many units, engineered and fitted together, result in a machine that facilitates transportation. A beautiful painting evinces design when paints of various colors are combined, by brush or knife upon canvas, so as to effect an esthetic response. Every intelligent person instinctively recognizes the presence of design.

> **"'Everything designed has a designer' is an analytically true statement."**
>
> — *Paul Ricci*

Could this happen by chance?!

The Miracle of Life

Multiplied thousands of examples of design are to be found in the various organisms of biological life that populate our planet. In this book, the primary point of focus will be upon that unique creature known as *Homo sapiens*, man, i.e., mankind.

Before giving consideration to the human body as an argument for design, hence, a Designer, it is fitting that the traits of a living organism be delineated. What are those factors which distinguish the organic (living) from the inorganic (non-living)? What is the difference between a living creature and a lifeless lump of clay?

It is generally agreed by scientists that an object may be defined as living when: (1) It is capable of metabolism, that is, it receives and breaks down elements outside of itself for the production of energy. (2) It experiences true growth, i.e., the multiplication of cells. (3) It is able to reproduce itself in independent organisms that replicate the original type. (4) It exhibits responsiveness (i.e., it reacts to external stimuli). (5) It is capable of autonomous movement. An automobile moves, but it is propelled by forces exterior to itself; a living organism is able to locomote itself.

There is no scientific information which explains the presence of life on Earth in a naturalistic way. The well-known Law of Biogenesis argues that life derives only from previously existing life. The notion that life accidentally initiated itself eons ago (i.e., spontaneous generation) is totally without scientific basis, though it is widely advocated by evolutionists.

Professor Edwin Conklin of Princeton University compared the random origin of life to an explosion in a print shop producing an unabridged dictionary (1963, p 62). Sir Fred Hoyle, one of Great Britain's prominent scientists, has argued that the chance of higher life-forms emerging accidentally is comparable to the chance that a Boeing 747 jet could be assembled by a tornado sweeping through a junk yard (1981A, p 105).

Dr. Hoyle also likened the random construction of life to 10^{50} (one followed by fifty zeros) blind men simultaneously solving scrambled Rubic's cubes (1981B, pp 521-27). Evidence points to the fact that life cannot generate itself. It must be concluded that this phenomenon commenced as a result of a supernatural act of creation. For a good discussion of "spontaneous generation," see Dr. Bert Thompson's material (1986, pp 59-68).

The Biblical View

How did man's marvelous body come into being? Is he the consequence of blind, natural forces? Or has he been divinely designed by a Creator? The biblical writers take the view that the human body was fashioned by God (Genesis 2:7). In Psalm 139, David declared:

> "I will give thanks unto you [Jehovah]; for I am fearfully and wonderfully made. Wonderful are your works; and that my soul knows right well. My frame was not hidden from you, when I was made in secret, and curiously wrought in the lowest parts of the earth" (14-15).

I am fearfully and wonderfully made.

The expression "lowest parts of the earth" is an idiom for the womb. The psalmist thus described the intrauterine development of the human body. Of particular interest in this passage is the expression "curiously wrought." It derives from a Hebrew term which denotes that which is woven or embroidered. In Exodus 26:36 the word describes the beautifully embroidered curtain/ door of the tabernacle. In the context of Psalm 139, the term is "applied by a natural metaphor to the complex and intricate formation of the body" (Kirkpatrick, 1906, p 789). Derek Kidner notes that this passage is a reminder of the value that God places on us, "even as embryos" (1975, p 466).

There is an interesting passage in the New Testament that complements the affirmation of David. In his first epistle to the Corinthians, Paul encourages unity among the Christians in that city. The apostle uses the unity of the human body as an example of the type of oneness that should characterize the people of God. In that connection, Paul writes: "But now God set the members each one of them in the body, even as it pleased him" (1 Corinthians 12:18).

W. E. Vine observed that the aorist tense form of the verbs in the Greek New Testament "marks the formation of the human body in all its parts as a creative act at a single point of time, and contradicts the evolutionary theory of a gradual development from infinitesimal microcosms" (1951, p 173).

The Organization of the Body

Remember that our initial definition of "design" stressed the arrangement of multiple parts into an organized unit for the accomplishment of a specific purpose. That is precisely the nature of the human body. We would rarely agree with noted evolutionist George G. Simpson, but for once we must concur that in man one finds "the most highly endowed organization of matter that has yet appeared on the earth . . ." (1949, p 293).

For organizational purposes, the body may be considered at four levels. (1) The smallest unit of life within the body is the *cell* (from a Latin word meaning "room"). A cell is a microscopic unit of organized life. Cells come in different types, sizes, and shapes, depending upon the kind of work they were designed to do. (2) A group of the same kind of cells that carry on the same activity is called a *tissue*. There are several kinds of tissue in the body (e.g., muscle tissue, nerve tissue, etc.). (3) A group of different tissues, all working in unison, is called an *organ*. Organs, such as the heart, liver, eyes, etc., conduct special activities within the body. (4) A group of organs orchestrated so as to carry on a special bodily function is called a *system*.

There are some ten major systems within the body (e.g., the digestive system, the circulatory system, etc.). It is therefore quite clear, to any knowledgeable and clear-thinking person, that the physical body has been marvelously designed and intricately organized, for the purpose of facilitating human existence upon the planet Earth. In this volume, some of the various features of the human body will be considered as examples of design, which must obviously point to the grand Designer.

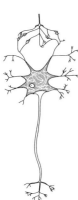

Nerve Cell

Design in the Cell

"The adult human body is estimated to contain 60,000 billion cells, every one of them subject to the rules and regulations of the group" (Pfeiffer, 1964, p 15). Who was the rule-Maker? [Note: Beck asserts that the human body contains 100 trillion cells (1971, p 189).]

Cells come in different sizes and shapes. On average, each of them is less than a thousandth of an inch in length. Some 40,000 red blood cells will fit into the letter O. "We have about a million cells in every square inch of our skin, and about thirty billion

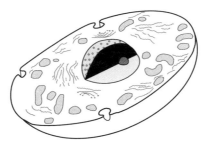

*". . . all cells are built according to a **fundamental deisgn**."*

in our brains" (Gore, 1976, p 358).

The shape of the cells is "related to their function; human red-blood cells are saucer-shaped and fairly flat, permitting the ready transfer of the oxygen and carbon dioxide they carry through the body, while nerve cells have long, thin extensions to transmit messages" (Pfeiffer, 1964, p 9).

Would anyone question the fact that the transmission features of a telephone system were designed? Why then would one deny the obvious design in the even more complex transmitting apparatus of the nerve cells? Pfeiffer admits:

> ". . . all cells are built according to a fundamental design which provides them with certain common features apparently necessary to life" (1964, p 10; emp. added).

Rick Gore describes the cell as a "microuniverse" which abounds with "discrete pieces of life, each performing with exquisite precision" (1976, p 358). He characterizes cell division as a process of "supreme design," and this evolutionist marvels at the "wisdom [that] is built into the cell's surface" (1976, p 373). A mere random occurrence? Never!

The cell may be studied under three major categories. First, there is the *cell membrane* which encloses the organism. Second, there is the watery *cytoplasm*, containing specialized features. This constitutes the bulk of the interior. Third, within the cytoplasm is the *nucleus*, the control center of the cell.

The cell membrane consists of very thin (about three-millionths of an inch) sandwich-like layers of protein and fat which form the outer protective coating of the cell. It is a semi-permeable, filter-like structure which allows only certain elements to enter or exit the cell.

Dr. Ernest Borek, a Professor of Microbiology at the University of Colorado School of Medicine, an evolutionist mind you, described the cell membrane as follows:

> "The membrane recognizes with its uncanny molecular memory the hundreds of compounds swimming around it and permits or denies passage according to the cell's requirements" (1973, p 5).

Surely this "uncanny molecular memory" must have been planned by a Mind!

There are a number of specialized components within the cell's cytoplasm. For example there are mitochondria, which are "miniature power plants" (about 1,000 in each cell), burning the food taken in, thus providing the cell with energy. The cell has tiny networks known collectively as endoplasmic reticulum (ER for short), which are "believed to be a transport system *designed* to carry materials from one part of the cell to

A planned operation.

another" (Pfeiffer, 1964, p 13, emp. added).

Again, note Pfeiffer's use of the word "designed" — a slip no doubt! Also in the cell are microscopic units called ribosomes. These are little factories which manufacture protein. Pfeiffer characterizes the cooperative effort between the ribosomes and the ER as a "joint operation" between "manufacturing and trucking" firms (1964, p 22). And yet, all of this is supposed to have evolved purely by chance? Incredible!

Additionally, cells have bag-like structures called Golgi bodies. It is believed that the Golgi apparatus packages and stores proteins, which the cell "exports." There are also small organelles called lysosomes which, among other things, function as efficient garbage disposal units. Clearly, this mechanism evinces intelligent design.

The nucleus is the "brain" of the cell. It is separated from the cytoplasm by a nuclear membrane. Within the nucleus are chromosomes — long, threadlike bodies that consist of proteins and a chemical called DNA (deoxyribonucleic acid). DNA is a super-molecule that carries the genetic information necessary for the replication of the cell. In human beings, the strands of DNA in each cell, if unraveled, would be about six feet long, yet they are less than a trillionth of an inch thick (Weaver, 1984, p 822).

It is estimated that if all the DNA strands in the adult human were tied end-to-end, they would reach to the sun and back (186 million miles) 400 times. If decoded and translated into English, the DNA in a single human cell would fill a 1,000 volume set of encyclopedias of approximately 600 pages each (Gore, p 357). Yet, amazingly, all of the zygote DNA necessary to make every human being on earth today (more than 6 billion people) could be fitted into a container about the size of an aspirin tablet. Are we to believe that all of this information came into existence accidentally?

A DNA molecule is composed of units of nucleotides. These are chemical combinations of sugar-phosphate and four bases — adenine, thymine, guanine, or cytosine. These bases bond the nucleotides in the spiral DNA molecule. In a strand of DNA the nucleotides are arranged in a specific order, along what sort of looks like a twisted ladder. The order of arrangement forms the "blueprint" that regulates the production of all living things.

Atomic physicist George Gamow described the DNA code as a "well-planned structure in which each atom or atomic group sits in its predetermined place" (1966, p 264). The fascinating question is, who planned it?

It is interesting that whereas DNA is composed of the same constituents wherever it is found — in a maple tree, a mouse, or a man — the "program" in each case says, "Make a mouse, make a man, etc." Moreover, in each of the billions of cells within the human body, the entire blueprint for the whole person is contained; yet, amazingly, each cell has been engineered so as to make only a specific part of the body, such as the eye, bone, the liver, connective tissue, etc.

The set of genetic instructions for humans is roughly 3 billion letters long (Radman & Wagner, 1988, p 40). There are a couple of very important points that need to be made with reference to these data.

First, though the DNA contains a very definite code for the production of living things, the message *per se* does not reveal its origin. The DNA code has been compared to the information stored on the floppy disk of a computer, or in a computer microchip.

One writer, in describing how much more information a DNA molecule contains than a much larger microchip, says:

> "We marvel at the feats of memory and transcription accomplished by computer microchips, but these are gargantuan [in size] compared to the protein granules of deoxyribonucleic acid, DNA" (Block, 1980, p 52).

The important point here is this: a programmed message is not self-explanatory in terms of its origin. One must assume that someone wrote the initial program. A program does not write itself! Similarly, it is obvious that Someone has programmed the data in DNA.

In their highly acclaimed book, *The Mystery of Life's Origin*, Thaxton, Bradley, and Olsen raise this interesting question:

> ". . . an intelligible communication via radio signal from some distant galaxy would be widely hailed as evidence of an intelligent source. Why then doesn't the message sequence on the DNA molecule also constitute *prima facie* evidence for an intelligent source?" (1984, p 211).

Dr. James Coppedge expressed the matter like this:

> "By all the rules of reason, could there be a code which carries a message without someone originating that code? It would seem self-evident that any such complex message system, which is seen to be wise and efficient, requires not only an *intelligence* but a *person* back of it" (1973, p 138; emp. in the original).

Second, this observation should be noted. Even though DNA contains the code of life, it is unable to directly implement the code into the production of tissue. Another substance, RNA (ribonucleic acid), accomplishes this. Thus, DNA and RNA work together to assemble the human body.

This analogy may help. DNA is the designer of the body, but it does not assemble the product. DNA creates an acid called RNA to do that job. It is like an architect designing a house and then turning the blueprint over to a carpenter to do the construction. Again, it must be stressed — the cooperative labor of these components argues very eloquently for design, hence, a Designer!

Conclusion

I want to conclude this chapter by calling attention to a fundamental form of logical argument that is called the *a fortiori* principle. This principle attempts to show that what is demonstrably true in one instance is even more likely to be true in another.

Here is an example: both a pair of pliers and a computer are tools. If one assumes that it took a designer to make the pliers, it surely will follow that it took a designer to make the computer, since the computer is much more complicated than the pliers. That is simple logic.

With this principle in mind, please examine the following quotations with reference to the living cell. Miller and Goode, two evolutionists, have written:

> "The cell has been likened to a power plant, a furnace, a chemical laboratory. In its reproductive functions it has been described as a factory complete with manager's office, files of blueprints and plans, intercommunication system, assembly line with foremen and workers.

> "None of these fanciful analogies does justice to the living cell. All of these man-made systems put together, however ingenious and efficient, could not reproduce the functioning of this single unit of life, too small to be seen with the unaided eye" (1960, p 162).

If the living cell is more "ingenious" than any "man-made" system, who made it? Are we to conclude that it just happened? That is wholly illogical.

Note this quotation from the *World Book Encyclopedia*:

> "... a cell can be thought of as a tiny chemical factory. It has a control center that tells it what to do and when. It has power plants for generating energy, and it has machinery for making its products or performing its services" (Rubenstein, 1979, 3: 250b).

Do factories happen by chance? Pfeiffer said that the cell "utilizes a tightly organized system of parts that is much like a tiny industrial complex. It has a central control point, power plants, internal communications, construction and manufacturing elements" (1964, p 16).

Surely these quotations are inadvertent concessions that the living system must have had a Designer. Significantly, professor William S. Beck of Harvard University, an evolutionist, has authored a book titled, *Human Design* (1971), though he obviously did not accept the logical conclusion of that appellation. The human body is not a fortuitous accident birthed by that mythical lady, "mother nature." Rather, "It is HE that hath made us" (Psalm 100:3).

Endnotes

Beck, William S. (1971), *Human Design* (New York: Harcourt, Brace, Jovanovich, Inc.).

Block, Irvin (1980), *Science Digest*, September/October - Special Issue.

Borek, Ernest (1973), *The Sculpture of Life* (New York: Columbia University Press).

Conklin, Edwin (1963), *Reader's Digest*, January.

Coppedge, James F. (1973), *Evolution: Possible or Impossible?* (Grand Rapids: Zondervan).

Gamow, George (1964), *One, Two, Three . . . Infinity* (New York: Viking Press).

Gore, Rick (1976), *National Geographic*, September.

Hoyle, Fred (1981A), *Nature*, November 12.

Hoyle, Fred (1981B), *New Scientist*, November 19.

Kidner, Derek (1975), *Psalms 73-150* (London: Inter-Varsity Press).

Kirkpatrick, A. F. (1906), *The Psalms* (Cambridge University Press).

Miller, Benjamin and Goode, Ruth (1960), *Man and His Body* (New York: Simon and Shuster).

Pfeiffer, John (1964), *The Cell* (New York: Time Inc.).

Radman, Miroslav and Wagner, Robert (1988), *Scientific American*, August.

Ricci, Paul (1986), *Fundamentals of Critical Thinking* (Lexington, MA: Ginn Press).

Rubenstein, Irwin (1979), in: *World Book Encyclopedia* (Chicago: World Book - Childcraft International, Inc.).

Simpson, George G. (1949), *The Meaning of Evolution* (New Haven: Yale University Press).

Thaxton, Charles B., Bradley, Walter L., Olsen, Roger L. (1984), *The Mystery of Life's Origin* (New York: Philosophical Library).

Thompson, Bert and Jackson, Wayne (1986), *Essays In Apologetics, Vol. II* (Montgomery, AL: Apologetics Press).

Vine, W. E. (1951), *First Corinthians* (Grand Rapids: Zondervan).

Weaver, Robert F. (1984), *National Geographic*, December.

CHAPTER 2

David, king of Israel, confidently declared: "I am fearfully and wonderfully made" (Psalm 139:14). Three thousand years of medical advancement have demonstrated the amazing accuracy of this affirmation.

A recent book published by the Reader's Digest Association declares:

> "When you come right down to it, the most incredible creation in the universe is you — with your fantastic senses and strengths, your ingenious defense systems, and mental capabilities so great you can never use them to the fullest. Your body is a structural masterpiece more amazing than science fiction" (Guinness, 1987, p 5).

In the previous chapter, we commenced the presentation of an argument that develops as follows:

(1) If it is the case that an object evinces design, then it must have had a designer.

(2) But it is the case that the human body evinces design.

(3) Thus, it is the case that the human body must have had a designer.

The reader should briefly review the preceding chapter to refresh his mind with the information as to what constitutes "design."

"Your body is a structural masterpiece more amazing than science fiction."

— *Reader's Digest Association*

As mentioned earlier, the human body can be studied at four major levels — cell, tissue, organ, and system. All of these are highly integrated arrangements within the physical frame. In this discussion consideration will be given to the design characteristic of some of the major systems of the human anatomy.

The Skin System

The skin system consists of three areas: the skin layers, the glands, and the hair and nails. Each of these is characterized by marvelous qualities of obvious design.

The skin itself is the largest organ of the human body. If the skin of an average 150 pound man were spread out, it would cover 20 square feet of space and weigh about 9 pounds. The skin is a busy area. "A piece of skin the size of a quarter contains 1 yard of blood vessels, 4 yards of nerves, 25 nerve ends, 100 sweat glands, and more than 3 million cells" (Youmans, 1979, 17: 404d). These numbers will vary at different body locations.

The skin, containing two major layers, is, on average, only about 1/8 of an inch thick. The epidermis is the top segment. It consists of rows of cells, about 12 to 15 deep. The upper layers are dead and are constantly being replaced by newly formed cells. *What man-made house replaces its own covering?*

The epidermis contains a pigment called melanin, which gives the skin its color. The underneath layer of skin is the dermis. It is joined to the epidermis by a corrugated surface which contains nerves and blood vessels. When a cut finger draws blood, the dermis has been reached. The dermis contains two kinds of glands — sweat and oil.

The ends of the fingers and toes are protected by a horn-like substance called nail. Actually, most of the nail that you see is dead; only the lower, crescent-shaped, white portion is living. The fingernails grow about three times as fast as the toe-nails. Certainly there is design in this, due to the respective functions of the hands and feet.

Designed for gripping.

The skin of the underside of the fingers, the palms, and the soles of the feet have a special friction surface. This area has no hair and, like the knurling on a tool handle or the tread of a tire, it has been designed for gripping (Miller & Goode, 1960, p 345).

Hair has several functions. It is a part of the skin's sentry system. Eyelashes warn the eyes to close when dust strikes them. Body hairs also serve as levers, connected to muscles, to help squeeze the oil glands. Hair acts as a filter in the nose and ears. Hair grows to a certain length, falls out, and then, in most instances, is replaced by new hair. Hair is "programmed" to grow a certain length. Eyelashes obviously do not grow as long as scalp hair. Who planned it this way? Clearly there is design in this circumstance.

Compared to most mammals, man is relatively hairless. Why is this? A strong case can be made for the fact that the best explanation is to be found "in the design of the human body with personhood in view" (Cosgrove, 1987, p 54). Skin touch is very closely associated with human emotions.

Human skin is one of the body's most vital organs. Its value may be summarized as follows:

(1) The skin is a protective fortification which keeps harmful bacteria from entering the human system.

(2) It is a waterproof wall which holds in the fluids of the body. Our bodies are about 75% fluid.

(3) It protects the interior parts of the body from cuts, bruises, etc.

(4) With its pigment, it shields the body from harmful rays of the Sun. Beck calls melanin "an epidermal light filter" (1971, p 745). Do light filters invented by man require intelligence?

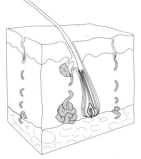

What man-made machine lubricates itself?

(5) The skin's many nerve endings make it sensitive to touch, cold, heat, pain, and pressure. It is thus a major sense organ.

(6) The sweat glands (2 to 5 million in the whole body) help eliminate waste products and also function in cooling the skin.

(7) The oil glands lubricate the skin and keep it soft — at the same time, waterproofing it. Though soft, the

skin is quite durable. When a 2,000 - year - old Egyptian mummy was fingerprinted, the ridges were found to be perfectly preserved (Guinness, 1987, p 132).

(8) About 1/3 of the body's blood circulates through the skin. The blood vessels, by contracting and expanding, work to regulate body temperature. If body temperature increases by 7 or 8 degrees fahrenheit and remains there for any length of time, a person will almost always die. The skin is thus a radiator system (Brand & Yancey, 1980, p 154). Does a radiator happen by accident?

(9) The skin absorbs ultraviolet rays from the Sun and uses them to convert chemicals into vitamin D, which the body needs for the utilization of calcium. The skin is a chemical-processing plant.

The Skeletal System

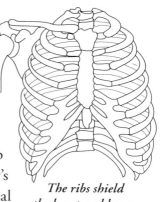

The ribs shield the heart and lungs.

The average adult has 206 bones in his body (an infant has about 350, many of which fuse during the maturation process). The human skeleton makes up about 15-20% of the body's weight. Bones serve several vital functions in the body

(1) They have been designed to be a rigid support for the organs and tissues of the body. Bones are like the interior framework of a house. The skeleton system is "something of an *engineering marvel*, strong enough to support weight and carry burdens, yet flexible to cushion

shocks and allow for an extraordinary variety of motion" (Miller & Goode, 1960, p 25, emp. added). Who was the engineer?

(2) Bones function as protective devices for many of the softer parts of the anatomy. For example, certain sections of the skull, which are independent in infancy but have grown together in the adult, offer protection for the fragile brain. The 12 pairs of ribs form a cage to shield the heart and lungs. The backbone (called the spinal column) is made up of 33 block-like bones which are ingeniously designed to allow movement, yet these bones protect a major feature of the nervous system—the spinal cord.

(3) Bones also serve as levers. Miller and Goode comment:

> "When our muscles move us about, they do it by working a series of articulated levers that make a most efficient use of every ounce of muscular motive power. The levers are the bones of the body's framework, fitted together with the neatness of jigsaw pieces and hinged by joints that must win the admiration of any mechanic" (1960, p 25).

Again, we must remind ourselves that these writers are evolutionists, not creationists.

(4) Bones also have a metabolic function. Until fairly recently it was assumed that bones were inert tissue, but studies have revealed that they are "constantly being remodeled" (Beck, 1971, p 626). They provide a reservoir of essential minerals (99% of the calcium and 88% of the phosphorus, plus other trace elements) which must

be rebuilt continuously.

Consider this: without calcium, impulses could not travel along the nerves, and blood would not clot. The interrelation between body-systems is phenomenal.

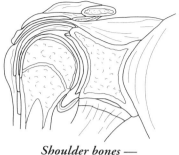

Shoulder bones —
what engineer designed these?

Too, red blood cells (180 million of which die each minute), certain white blood cells, and platelets arise in the marrow of the bones. Incredibly, when a bone is broken it immediately begins to repair itself; after the repair, it will be even stronger than before!

> "Perhaps an engineer will someday develop a substance as strong and light and efficient as bone, but what engineer could devise a substance that, like bone, can grow continuously, lubricate itself, require no shutdown time, and repair itself when damage occurs?" (Brand & Yancey, 1980, p 91).

We will leave it to the evolutionists to figure out how that "nature," poor, blind nature, with no intelligence, just "thought up" this process!

In order for a skeletal system to be effective, it must have several attributes: strength, elasticity, and lightness of weight. Amazingly, Someone designed the bones with all of these characteristics. Bones are very strong. A cube of bone 1 square inch in surface will bear, without being crushed, a weight of more than 4 tons. Ounce for ounce, bone is stronger than solid steel. And yet, a piece of bone will stretch 10 times as much as steel. A steel frame comparable to the human skeleton would weigh 3 times as much.

Dr. Alexander Macalister, a former professor of anatomy at Cambridge University, stated:

> "Man's body is a machine formed for doing work. Its framework is the most suitable that could be devised in material, structure, and arrangement" (1886, 7:2).

As a specific example of bone design, consider the bones of the foot. One-fourth of all the body's bones are in the feet. Each human foot contains 26 bones. The feet have been ingeniously designed to facilitate a number of mechanical functions. They support, with arches comparable to an engineered bridge. They operate as levers when one presses an automobile accelerator peddle. Feet act like hydraulic jacks when one tip-toes. They catapult a person as he jumps. And feet act as a cushion for the legs when one is running. All of these features are quite helpful — especially in view of the fact that an average person will walk about 65,000 miles in his/her lifetime — equivalent to traveling around the world more than 2 1/2 times. The skeletal system demonstrates design; there must have been a brilliant Designer. There was — God.

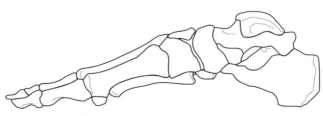

The foot — a well-designed arch!

The Muscle System

There are more than 600 muscles (containing about 6 billion muscle fibers) in the human body, making up about 40% of the body's weight. According to Dr. I. MacKay Murray, professor of anatomy at State University of New York, the muscles are the "engines" of the body, providing the power for movement (1969, p 22). Do "engines" develop accidentally? Some muscles are tiny, such as those regulating the amount of light entering the eye, while others, like those in the legs, are massive.

Muscles are classified as "voluntary" (under the control of the will) or "involuntary" (not under mind control). The voluntary muscles of the arms, for example, are connected to the bones by tough cords of connective tissue called tendons. One must "think" to move these muscles. The involuntary muscles are those whose contraction and relaxation cannot be consciously controlled, e.g., the heart and the intestines. [Note: The heart is a unique type of muscle and is usually considered in connection with the cardiovascular system.]

All muscles, in one way or another, are regulated by the nervous system. Some muscles are both voluntary and involuntary. The muscles that control the eyelids and the diaphragm (for breathing) are examples of these.

Muscles work by tightening or contracting. When they contract, they shorten, thus exerting a "pull"; muscles do not "push." Frequently muscles work in pairs, as in the voluntary skeletal muscles. The biceps in the upper arm pulls the forearm upward, whereas the triceps moves the forearm downward. While one works, the other rests; the design is amazing.

Some muscles, like those attached to the skeleton, are analogous to strong steel cables. Each muscle is constructed of long cells combined in small bundles called fibers. These bundles are bound together, making larger bundles of which the whole muscle consists. Muscle fibers vary in size from a few hundred-thousandths of an inch, to an inch or inch-and-a-half in length.

**"The body's engines [muscles] . . .
demonstrate some
surprisingly modern engineering ideas."**

— Dr. John Lenihan

Each muscle has its own stored supply of high-grade fuel, especially sugar (glycogen) which the body has manufactured from food that has been consumed. This analogy may be helpful. In the automobile engine, the spark ignites vaporized gasoline, the piston moves, and keeps moving in response to a series of explosions.

> "A muscle cell performs the functions of both the spark and the piston; the cell itself splits a molecule of fuel and also exerts the resulting physical power" (Miller & Goode, 1960, p 23).

If it is clear that the automobile engine was intelligently designed, why is it not reasonable to draw the same conclusion with reference to the muscles? Dr. Lenihan, even though an evolutionist, writes: "The body's engines [muscles - WJ] . . . demonstrate some surprisingly modern *engineering* ideas" (1974, p 43, emp. added). The question is, who initiated these ideas? The answer is, our Maker did.

Connected to the skeletal muscle is a nerve. The nerve conveys a signal telling the muscle when to contract or relax. Obviously there must be precise orchestration between the skeletal muscle system and the nervous system. Without doubt, their cooperative nature was planned. Some muscles, like those in the stomach, are stimulated to work by means of chemicals called hormones.

There is also a precisely integrated relationship between muscles and bones. Here is one example. "As certain muscles increase in strength, they pull harder than before on the bones to which they are attached. With this as a stimulus, bone-forming cells build new bone to give internal reinforcement where necessary" (Shryock, 1968, p 27). Would this indicate planning?

In his book, *Human Design*, evolutionist William Beck could hardly contain himself when he wrote of "the intricate structural organization" of the muscles and tendons in the hand, which are capable of such a wide variety of actions. Remember, "intricate structural organization" indicates design. He characterized this phenomenon as "one of evolution's most remarkable achievements" (1971, p 691). Remarkable indeed!

It is the epitome of gullibility to think that this ingenious device came about merely by a series of quirks in nature. An essay on the human hand appeared some years back in the magazine, *Today's Health* (published by the American Medical Association). Though saturated with evolutionary propaganda (e.g., the hand evolved from a fish's fin), the article conceded:

> ". . . If the most gifted scientists cudgeled their brains they probably could not come up with a stronger or more perfect tool for grasping and delicate manipulation than the human hand. And seen from an *engineering* standpoint, the loveliest hand actually is a highly complex mechanical device composed of muscle, bone, tendon, fat, and extremely sensitive nerve fibers, capable of performing thousands of jobs with precision" (Wylie, 1962, p 25, emp. added).

Something "engineered" requires an engineer; that's just sound logic.

While many living organisms share common muscle activity, there are some muscle movements that are unique to man. These forcefully demonstrate that the human being is not an evolved animal; rather, he is a creature "fearfully and wonderfully made" by a Creator.

Observe the following quotation from two evolutionists, which certainly reveals more than these learned authors intended. And then, ask yourself: how can scientists echo these sentiments and still cling to the evolution-

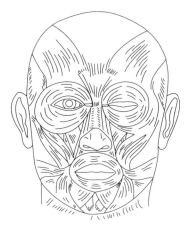

*Our facial muscles provide us with the most
intricate system of facial expression.*

ary dogma? One is almost forced to the conclusion that they reflect a religious bias, rather than scientific objectivity.

> "Only man can combine muscle with intelligence and imagination, plan and purpose, to plow and plant a field, to create a museum masterpiece or the 'Gettysburg Address.' And only man trains to perform the most highly co-ordinated forms of bodily motion for their own sake, in the expressive and athletic arts. We applaud this skill in our species every time we clap our hands for a ballerina or a circus aerialist" (Miller & Goode, 1960, p 21).

In this connection there is another point of interest. "Of all the creatures in the world, the human being is endowed with the most complex face to present to the world. Our facial muscles provide us with the most intricate system of facial expression" (Cosgrove, 1987, pp 24, 25). There are about 28 muscles which are involved in making the various facial expressions of which human beings are capable. With facial muscles we can reveal anger, surprise, perplexity, amusement, joy, etc. It is possible for humans to make a quarter of a million different facial changes. It is important to observe that only human beings are able to smile. [Note: Primates may appear to be smiling when their lips are drawn back with their teeth showing, but this is actually an expression of anxiety (Cosgrove, 1987, p 34).]

A baby will spontaneously smile a few hours after birth. Significant, however, is the fact that unless smiles are reinforced by talking, touching, etc., they soon disappear. Again, it must be stressed, the muscular design of the human face is quite unique. Even evolutionists admit:

> "Many animals can produce a grimace or a snarl. But only man is equipped with such an exquisitely differentiated set of muscles — the mimetic musculature of the face — with no other function than to express and communicate feelings" (Miller & Goode, 1960, p 22).

Conclusion

In this chapter we have only discussed three of the ten major systems of the human body. A fair consideration of any one of these would lead only to the conclusion that mankind was intelligently fashioned by a supreme Mind.

As we emphasized earlier, if obvious design is recognized in the man-made world, why should it not be acknowledged in the natural world?

A prestigious science journal suggested:

"A pair of pliers, a chain saw or even a missile guidance system doesn't approach the lowliest parasitic worm in internal complexity. The human-made world is not nearly as intricate as the natural world" (Mankin, 1981, p 18).

Also reflect upon this quotation from a trio of militant evolutionists:

"A modern building is certainly a complex and highly ordered structure, but its complexity cannot begin to compare with that of the living system. And for precisely the same reasons that make us reject the idea of a building coming into existence spontaneously, we are forced to reject the idea that anything as complex as an organism could arise spontaneously from the materials of the nonliving world" (Simpson, Pittendrigh, Tiffany, 1957, p 262).

No one, aware of the facts, would accuse George G. Simpson and his colleagues of religious bias! Or hear the testimony of Dr. Lenihan: ". . . the body is vastly more complicated than any man-made engine" (1974, p 152).

It is thus unreasonable to contend that "man the machine" is merely the fortuitous result of a series of millions of "happy accidents," to use the expression of Simpson and his co-authors (1957, p 451). That is a fairy tale for those who refuse to have the Creator in their knowledge (Romans 1:28).

As we reflect upon the various elements of design that are so apparent in the human body, it is important that we stress that we are not calling upon our readers to draw a fanatical, emotional conclusion with reference to these matters. We are simply suggesting that if one uses his mind — if he thinks logically, he will be forced to abandon the notion that the features of the human physical system could have formed gradually and accidentally over a vast era of time. This is what the evolutionary theory insists, but it does not conform to the facts.

As we consider the brilliant design that is evident in the body, we should ask: Why were we thus designed? There is little value in acknowledging that a Creator was responsible for our existence if we do not want to know the reason for our presence upon the Earth, and thus, what responsibility we sustain to Him.

This answer to this fascinating inquiry will not be found in the medical laboratory. It is in the Bible.

". . . the body is vastly more complicated than any man-made engine."

— Dr. John Lenihan

Endnotes

Brand, Paul and Yancey, Phillip (1980), *Fearfully & Wonderfully Made* (Grand Rapids: Zondervan).

Beck, William S. (1971), *Human Design* (New York: Harcourt, Brace, Jovanovich).

Cosgrove, Mark P. (1987), *The Amazing Body Human* (Grand Rapids: Baker Book House).

Guinness, Alma E., Ed. (1987), *ABC's of The Human Body* (Pleasantville, NY: Reader's Digest Association).

Lenihan, John (1974), *Human Engineering* (New York: George Braziller, Inc.).

Macalister, Alexander (1886), "Man Physiologically Considered," *Living*

Papers, Vol. VII (Cincinnati: Cranston & Stowe).

Mankin, Erik (1981), "Darwin and the Machines," *Science Digest*, 89[3]:18.

Murray, I. MacKay (1969), *Human Anatomy Made Simple* (Garden City, NY: Doubleday & Co., Inc.).

Miller, Benjamin and Goode, Ruth (1960), *Man and His Body* (New York: Simon and Schuster).

Science Digest (1981), April.

Shryock, Harold (1968), "Your Bones Are Alive!," *Signs of the Times*, January.

Simpson, George G., Pittendrigh, C. S., Tiffany L. H. (1957), *Life: An Introduction to Biology* (New York: Harcourt, Brace and Company).

Wylie, Evan McLeod (1962), *Today's Health*, July.

Youmans, W. B. (1979), in:*World Book Encyclopedia* (Chicago: World Book-Childcraft International, Inc.).

CHAPTER 3

Philip Johnson is a Harvard-trained attorney who is currently a professor at the University of California (Berkeley). Professor Johnson has authored a controversial book, *Darwin on Trial,* in which he argues that evidence for proving the theory of evolution simply does not exist. Johnson's book was recently reviewed by Dr. Arthur Quinn, a colleague at the University. Quinn, who alleges that "Darwinism has been a highly productive theory in biology" (1992), nevertheless makes the following startling confession:

> "Common sense seems initially on the side of creationism. The intricate interrelations of parts of organisms and their beautiful adaptations to their environment does seem indisputable evidence of the handiwork of a benign power" (1991, p 23).

Sadly evolutionists lay "common sense" aside, and accept the Darwinian concept for a variety of other reasons. Note, however, the professor's recognition that "intricate interrelations" appear to suggest a Creator.

In earlier chapters, we have argued for the existence of a Creator upon the basis of the Law of Teleology, namely, where there is evidence of design, there must be a designer — a principle conceded even by skeptics. In our discussion of the nature of teleology, we noted that *when an organism is characterized by intricate interrelations, it becomes clear that it has been designed.* (See quote above.)

Our argument has been developed especially with

"Common sense seems initially on the side of creationism."

— *Prof. Arthur Quinn*

---

reference to the human body. We have reasoned in this fashion: If it is the case that the human body evinces design, then it must have had a designer. But it is the case that the human body evinces design. Therefore, it is the case that the human body must have had a designer. Since the major premise is not disputed, our task has been to introduce evidence which argues that the human body bears the marks of design.

We have previously discussed the design that is apparent in: (1) the living cell; (2) the skin system; (3) the skeletal system; (4) the muscular system. In this discussion, consideration will be given to two other systems of man's body.

The Digestive System

For the human body to live it must have an energy supply. The food we eat supplies that energy. The average person will consume some 40 tons of food during a lifetime. How does the body process that food so that the energy is utilized? This is accomplished by a complicated network known as the digestive system.

The digestive system consists of two main areas: the alimentary canal and the digestive glands. The alimentary canal is the tube through which the food passes in the body. It includes such areas as: the mouth, throat, esophagus, stomach, and the small and large intestines. The digestive glands include: the salivary glands, liver, pancreas, the gastric glands of the stomach, and certain glands of the small intestine.

Dr. John Lenihan, who has been described as one of the world's most eminent bioengineers, compares the body's utilization of energy to the engine in an automobile. He writes:

> "It includes systems which can be recognized as a fuel tank, a carburetor, an exhaust pipe and even a supercharger — but the body is vastly more complicated than any man-made engine" (1974, p 152).

It is difficult to see how such scientists remain evolutionists, arguing that unintelligent forces in nature "invented" this amazing machine! If the body is "vastly more complicated" than the obviously designed automobile engine, logic clearly demands that it is not an accident.

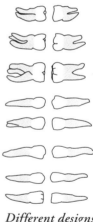

Food is taken into the mouth where it is chewed by the teeth, 32 of them in the adult. The teeth come in different shapes, adapted to their purposes. The

Different designs in teeth.

sharp incisors cut, the pointed canines were fashioned to rip or tear, while the premolars and molars are obviously designed for grinding. The teeth are covered by enamel — the hardest material in the body — which is resistant to chemicals that are potent enough to digest food.

In the mouth, food is introduced to the tongue.

The tongue is a marvelously complex instrument with several functions. It is a sense organ, warning, for instance, when food is too hot. The tongue manipulates the food, fashioning it into a small ball (bolus) which facilitates swallowing.

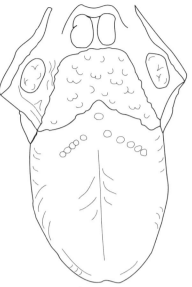

Most of the taste buds (about 9,000 in man) are located on the tongue thus enabling one to differentiate between tastes

The tongue is a complicated sense organ.

that are sweet, bitter, salty, or sour. Dr. Lenihan notes that the phenomenon of taste is the result of "sophisticated chemical technology" (1974, p 110). Can there be sophisticated technology without a technologist?

There is an interesting sidepoint here. It is problematical to evolutionists that whereas both man and the hare each have about 9,000 taste buds, the rabbit — presumably much more closely related to the hare than to man — has some 17,000 taste buds. This presents a difficulty for the evolutionist's "comparative anatomy" argument, i.e., the idea that the closer organisms are related, the more similar they will be.

In the mouth, food is bathed in saliva. More than a quart of saliva is secreted each day by three pairs of salivary glands. This liquid accomplishes several important functions. It moistens the food so that it can be molded into a form that is easy to swallow; additionally, its mucus lubricates the food to accommodate this process. Saliva contains an enzyme, ptyalin or amylase, which begins the conversion of starches into sugar. Finally, this chemical substance contains a bactericide, lysozyme, which is a very effective germ-killer. The old saying, "That's not worth spit," is not only crude, it is quite inaccurate as well.

Swallowing is an extremely intricate process. Evolutionist William S. Beck, in his textbook, *Human Design*, describes the "highly specialized musculature" function of swallowing as "quite complex" (1971, p 518). While the food is passing from the mouth into the esophagus (a 10 inch muscular tube leading to the stomach), passageways which connect up to the nose, and down to the lungs, must be closed off. If it were not for the highly orchestrated arrangement of muscles (controlled by the nervous system), one would suffocate while attempting to eat. How could man have survived while "nature" was perfecting this mechanism?

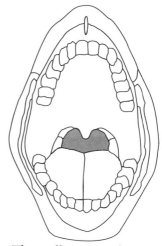

Food is pushed along the esophagus toward the stomach by a series of peristaltic movements. These are periodic muscular contractions. The stomach is a bag-like structure that serves as a temporary receptacle for food. It prepares the food for eventual treat-

The swallowing mechanism had to work right the first time!

ment by the small intestine. The average adult stomach can hold about 1½ quarts of food, which it retains for 3 to 4 hours.

During this time, the food is bathed with gastric juices which flow from three types of glands located in the wall of the stomach. The stomach is a truly remarkable structure. It is able to digest materials which are compositionally much tougher than it is.

> "We would have to boil our food in strong acids at 212° Fahrenheit to do with cookery what the stomach and intestines do at the body's normal temperature of 98.6°" (Miller & Goode, 1960, p 108).

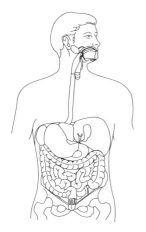

Another incredible thing about the stomach is the fact that though it consists of flesh, it does not digest itself! One of the chemicals in the stomach is hydrochloric acid. Hydrochloric acid is strong enough to dissolve a razor blade, and yet, under normal conditions, it does not harm the stomach. Why not? Scientists believe there are probably several factors involved.

> "First, the gastric lining is coated with mucus, which forms a barrier between the acid and the stomach wall. The mucus, somewhat alkaline, neutralizes the acid and thus helps to keep the stomach from digesting itself. Furthermore, food in the stomach dilutes the acid, making it less corrosive. Also, the lining of the stomach sheds cells at the rate of half a million every minute and replaces them so rapidly that the stomach has what amounts to a new lining every three days" (Guinness, 1987, p 242).

The stomach, with its various enzymes, breaks down and churns (with muscular movements) its contents into a semiliquid state called chyme. It might be noted that both the secretions and motor activity of the stomach are controlled by the brain via the nerve network. Without the cooperative

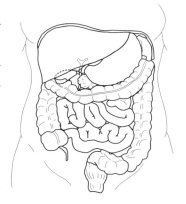

The Intestines

harmony between the digestive system and the nervous system, neither could function. This is more evidence of design.

The chyme is released into the small intestine where most of the digestive process takes place. The small intestine is a coiled tube some 20 to 25 feet long, and an inch to an inch-and-a-half in diameter. In the upper portion of the small intestine, the chyme is treated by digestive juices from other organs (e.g., the liver, pancreas, and gallbladder). Proteins, carbohydrates, and some fats are broken down and absorbed through the lining of the intestine. This nourishment is carried to all parts of the body by the blood. Waste products move into the large intestine for eventual elimination. The large intestine is about 5 feet long and 2 inches thick. The large intestine not only stores wastes, it also conserves water, and functions "as an incubator for a variety of bacteria that serve the body's nutritional needs" (Beck, 1971, p 550).

The digestive glands are a very essential part of the body's food-processing operation and should not be overlooked as thrilling examples of purposeful planning. For instance, the liver is the largest internal organ of the body. It weighs about 3 pounds. Amazingly, it performs about 500 different functions! It is such a complicated "chemical factory" that scientists do not believe that any humanly-devised machine could do its work (Guinness, 1987 p 244). The liver stores vitamins, detoxifies poisons, stabilizes the body's blood-sugar level, builds enzymes, etc. The liver filters enough blood in a single year to fill 23 milk trucks.

Close to the small intestine is the pancreas. Every day it secretes more than a pint of "juice" into the upper portion of the small intestine. Curiously, this small organ contains two kinds of cells. Some of these produce enzymes which facilitate the digestion of proteins, carbohydrates, and fats. Other cells produce insulin, which regulates the sugar economy of the blood. The pancreas is thus viewed as a part of the digestive system, and also a component of the endocrine system. The digestive system is a phenomenal example of design in the body. Clearly man is the product of a Maker!

The Circulatory System

The circulatory system consists of: the heart, the blood, and the blood vessels. This feature of the body has several important functions: (1) it transports digested food particles to the various parts of the body; (2) it takes oxygen to the cells for burning the food, thus, producing heat and energy; and, (3) it picks up waste materials and carries these to those "disposal" organs which eliminate wastes from the body.

The heart is a small muscle (some say two muscles connected) in the upper chest cavity. Dr. Michael DeBakey calls it a "busy machine" that pumps blood to all parts of the body (1984, 9:132a). Does a machine happen by accident?

In the adult male, the heart weighs about 11 ounces and is about the size of a large

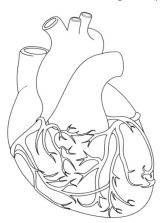

The Amazing Pump

fist; a woman's heart is slightly smaller. Miller and Goode describe this marvelous muscle as a "pump with a built-in motor" (1960, p 63). Is it not the case that something built has a builder?

The heart is the strongest muscle in the body. Normally it beats (in the adult) at about 70 to 80 times per minute. When the body needs an extra supply of blood (e.g., during vigorous exercise), it can beat 150 to 180 times a minute. This automatic regulating feature clearly indicates design. Note this unwitting testimony from a militant evolutionist.

> "The heart and blood vessels do more than speed or slow our blood flow to meet [the body's] needs. They carry the scarlet stream to different tissues under differing pressures to fuel different actions. Blood rushes to the stomach when we eat, to the lungs and muscles when we swim, to the brain when we read. To satisfy these changing metabolic needs, the cardiovascular system *integrates information as well as any computer, then responds as no computer can*" (Schiefelbein, 1986, p 124, emp. added).

The force the heart exerts is tremendous. It can squirt a stream of blood about 10 feet up into the air. In the span of 1 hour, the heart generates enough energy to lift a medium-sized car 3 feet off the ground (Avraham, 1989, p 13). The heart is an involuntary muscle that beats about 100,000 times a day, or nearly 40,000,000 times each year. It pumps about 1,800 gallons of blood a day. In a lifetime a heart will pump some 600,000 metric tons of blood!

To get some idea of the power of this muscle, try squeezing your fist tightly once each second for as long as you can. You will soon get tired and be unable to continue. Yet the heart goes on and on!

". . . it is hard to imagine a better job of engineering"

— *Miller and Goode*

What causes the heart to beat? It contains a small patch of tissue called the sinus node, or cardiac pacemaker. Somehow, about every 8/10 of a second, it produces an electrical current (a jump-start) to certain nerve fibers which stimulate the muscular contractions that send the blood flowing (at up to 10 miles per hour) throughout the body.

The body's blood supply, which gets depleted of oxygen, is pumped back to the heart. From there it is conveyed to the lungs where it is re-oxygenated and sent once more to the various parts of the body. Blood is thus being continuously pumped into, and out of, the heart with its rhythmic beating.

Evolutionists Miller and Goode cannot but concede that "for a pump that is keeping two separate circulatory systems going in perfect synchronization, it is hard to imagine a better job of *engineering*" (1960, p 68; emp. added).

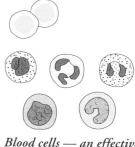

Blood cells — an effective transportation system.

And yet they believe that this amazing device, which they say is "hard to describe as anything short of a *miracle*" (p 64, emp. added), was engineered by the blind forces of nature. Incredible!

Medical authorities have observed that the heart's efficiency, i.e., the amount of useful work in relation to fuel used, is about twice that of a steam engine (Lenihan, 1974, p 131). If intelligence was required to invent the steam engine, does it not stand to reason that intelligence lies behind the more-efficient heart-pump?

Fifteen centuries before Christ was born, Moses declared that "the life of the flesh is in the blood" (Leviticus 17:11). This inspired truth was uttered more than 3,000 years before English physician William Harvey discovered the circulatory system in 1628.

Blood is actually classified as a tissue. The body contains about 5 to 6 quarts of this liquid tissue. The blood consists of plasma (which is mostly water), salts, a protein called fibrinogen, antibodies (which fight disease), enzymes, and hormones. The plasma helps maintain chemical balance in the body, regulates the body's water content, and assists in controlling temperature. The blood also contains solid materials — red cells, white

*The Amazing
Pipeline System*

cells, and platelets. The 25 trillion red cells transport oxygen throughout the body, and carry carbon dioxide back to the lungs (via the heart). The white cells (5 different kinds) attack bacteria and other germs. They are the body's defensive army. The platelets (15 million in each drop) help the blood to clot when the body is wounded. These are the body's repairmen.

Harmful bacteria and worn-out cells are filtered out of the blood by the liver and the spleen. The kidneys also remove waste products from the bloodstream. The blood has a very effective garbage disposal system.

Who could possibly believe that these wonderfully integrated mechanisms simply happened by chance? Atheists do!

In order for the blood to accomplish its vital work, it must remain at a relatively constant temperature. A radical drop in body temperature can damage the cells, and if the temperature rises above 108°F, one cannot long survive. Our Maker, therefore, placed a "thermostat" in the brain that monitors the temperature of the blood as it flows through that organ. When the air temperature drops, the heart slows down and the blood vessels constrict, forcing the liquid tissue to flow deeper within the body where it can remain warm. When the weather gets warm, or when we exercise, the arterioles open and the blood is dispersed within the skin, effectively functioning like a radiator (Schiefelbein, 1986, p 128). Surely such an orchestrated system cannot be explained in terms of a series of "happy accidents."

The blood vessels constitute an incredible pipeline system networking the entire body. These vessels come in three basic types. (1) Arteries (and smaller arterioles) are vessels that carry blood away from the heart. (2) Veins (and smaller venules) transport the blood back to the heart. (3) Capillaries are microscopic vessels that link the smallest arteries with veins.

"No engineering genius has ever invented a pump like the human heart."

— *Miller and Goode*

If all of the body's pipelines were connected end-to-end, it is estimated that it would stretch out from between 60,000 to 100,000 miles. The system is "so efficient" that the entire process of circulation, "during which every cell in the body is serviced, takes only a total of 20 seconds" (Avraham, 1989, p. 41). Would any rational person deny that a major city's pipeline system was carefully designed? Hardly. The body's skillfully constructed transportation system clearly evinces design, hence a Designer. Lenihan confesses: "The circulation is an example of a multipurpose system, often found in the body but generally *beyond the capability of the engineering designer*" (1974, p 5; emp. added).

In this connection it might be noted that medical scientists, interested in extending human longevity, have attempted to fashion numerous artificial organs. All such efforts have met with only limited success. One authority notes: ". . . no synthetic spare part — however well engineered — can match the capability of the organ a normal human being is born with" (Mader, 1979, p 367). Again, "No engineering genius has invented a pump like the human heart" (Miller & Goode, 1960, p 6).

Dr. Pierre Galletti of Brown Medical School describes artificial body-parts as "simplistic substitutes for their sophisticated natural counterparts" (see Cauwels, 1986, p ix). Man can attempt to duplicate God's handiwork, but he can never hope to approach the wisdom and skill of the Creator.

The arteries have been fashioned in a brilliant way so as to be both elastic and porous. The elasticity accommodates the surging blood, and also helps regulate body temperature. But how is the blood able to make its way, against gravity, back up the veins to the heart? They contain a series of one-way valves with open ends which face the heart — analogous to the valves in an automobile engine (Miller & Goode, 1960, p 71). The blood is partially pushed upward by force from the heart, but it is also propelled by muscle movements which massage the veins, pushing the blood forward through the valves. In the veins of the legs, these valves are spaced about every half-inch.

The capillaries are the smallest but most abundant blood vessels. It takes about 120 short capillaries to measure 3 inches. All of them laid end-to-end, however, would circle the equator twice (Avraham, 1989, p 40). The blood is pumped into the capillaries with a force sufficient to drive the plasma and its rich cargo through the porous walls of these tiny vessels thus nourishing the cells. This procedure requires a very "precise balance of pressures between the blood flowing within their walls and the fluid in and around the body's cells" (Schiefel-bein, 1986, p 114).

Without question this delicately balanced system affirms design. The master Engineer was God. Some struggle hard to deny this; at other times, they wonder:

> "If, like the scientists of an earlier day, we assumed a constant guiding *purposefulness* in our biological universe, we might say that the capillary system is the purpose of the circulation, that the entire system, heart and all, was designed for just this end" (Miller & Goode, 1960, p 77; emp. added).

How very sad. The human body reveals "design" in literally millions of ways.

As you will observe, over and over again we are using the word "design" in this book. We will not apologize for the repetitious use of the term. We will hammer the point home until an everlasting impression is made. That is what this book is all about — is the human body an accident, or was it *designed?* If it contains indications of design, we must confess, if we are candid, that there is a Designer. Let us honor Him, and be responsive to His will.

Endnotes

Avraham, Regina (1989), *The Circulatory System* (New York: Chelsea House Publishers).

Beck, William S. (1971), *Human Design* (New York: Harcourt, Brace, Jovanovich).

Cauwels, Janice (1986), *The Body Shop* (St. Louis, MO: The C. V. Mosby Co.).

DeBakey, Michael E. (1984), in: *World Book Encyclopedia* (Chicago: World Book – Childcraft International).

DuBos, Rene (1964) in: *The Cell* (author John Pfeiffer), "Introduction," (New York: Time, Inc.).

Guinness, Alma E., Ed. (1987), *ABC's of the Human Body* (Pleasantville, NY: Reader's Digest Association).

Lenihan, John (1974), *Human Engineering* (New York: John Braziller, Inc.).

Mader, Sylvia S. (1979), *Inquiry Into Life* (Dubuque, IA: Wm. C. Brown Co.).

Miller, Benjamin and Goode, Ruth (1960), *Man and His Body* (New York: Simon and Schuster).

Quinn, Arthur (1991), *California Monthly*, November.

Quinn, Arthur (1992), Letter to the author, January.

Schiefelbein, Susan (1986), in: *The Incredible Machine* (Washington, D.C.: National Geographic Society).

CHAPTER 4

Christians happily affirm it; skeptics begrudgingly concede it — logic. Everything designed had a designer. If, therefore, design is discernible in the Universe, there must have been a designer who fashioned it into its purposeful form.

The Scriptures declare that humanity is the product of design. David announced: "I am fearfully and wonderfully made" (Psalm 139:14). Man is not some fortuitous creature accidentally conceived by "father chance" and birthed by "mother nature." We are the offspring of God, in Whom we live, move, and have our very existence (Acts 17:28,29).

In earlier chapters, we have reasoned that the human body is a demonstration of design, hence, it points to a designer, namely God. In our argument we have called attention to the fact that the body is composed of a number of major of systems which are precisely integrated. Each system is dependent upon the others for its survival; none could function independently. These systems simply could not have evolved over eons of time. Moreover, the individual systems each reveal countless details of fascinating design that baffle even the brightest infidels.

**Man is not some fortuitous creature
accidentally conceived by "father chance"
and birthed by "mother nature."**

Dr. Alan Nourse, who reveals no religious bias, writes:

> ". . . the bodily mechanism is a masterpiece of precise planning, a delicate and complex apparatus whose various components work as a unit to achieve such diverse feats as scaling a mountaintop, building a bridge or composing a symphony" (1964, p 9).

A "masterpiece of precise planning"? Who planned it? And if it takes intelligence to build a bridge, or to compose a symphony, did it not require intelligence for the origin of the human body?

The Nervous System

The nervous system is the "communication center" of the body. It consists of: (1) the *brain;* (2) the *spinal cord;* and, (3) the *nerves,* which spread out from the brain and spinal cord to all parts of the body, somewhat like the root system of a tree.

The nervous system has many functions. It regulates the actions of organs like the muscles, liver, kidneys, etc. It monitors the senses, such as seeing, hearing, feeling, etc. The nervous system controls our thinking, learning, and memory capacity.

The nervous system is ". . . the most elaborate communications system ever devised."

— *Dr. John Pfeiffer*

The specialized nerve receptors in the sensory organs receive bits of information from the environment. For example, in the skin there are some 3 to 4 million structures sensitive to pain. There are a half million touch detectors and more than 200,000 temperature gauges. These tiny receptors, plus those in the eyes, ears, nose, tongue, etc., are constantly sending data to the brain. This information

Cervical nerves in the neck— more complex than a telephone system.

is transmitted (at up to 450 feet per second, or 300 mph), via the nerve fibers to the brain. The transmission involves both electrical and chemical energy. The brain analyzes the data and determines the appropriate action to be taken. Noted science writer John Pfeiffer, an evolutionist, calls the nervous system "the most elaborate communications system ever devised" (1961, p 4). Devised? Who devised it?

Several years ago the prestigious journal, *Natural History*, contained this statement: "The nervous system of a single starfish, with all its various nerve ganglia and fibers, is more complex than London's telephone exchange" (Burnett, 1961, p 17).

If that is true for the nervous system of the lowly starfish, what shall we say of the infinitely more complex human nervous system?

Again, "transmission of information within the nervous system is more complex than the largest telephone exchanges" (*Encyclopaedia Britannica*, 1989, 2:226).

If it took intelligence to design the telephone system, shall it be assumed that mere chance "designed" the more complicated human nervous system? That simply does not make sense.

The brain, located in the protective case called the skull, is the most highly specialized organ in the body. Isaac Asimov, well-known author, and radical humanist, said that man's brain is "the most complex and orderly arrangement of matter in the universe" (1970, p 10). Who arranged it? Paul Davies, professor of mathematics and physics at the University of Adelaide, declares that the human brain is "the most developed and complex system known to science" (1992, 14[5]4).

The human brain, which weighs about three pounds, consists of three main areas. The cerebrum is the thinking-learning center. It deciphers messages from the sensory organs, and controls the voluntary muscles. Evolutionist William Beck speaks of the "architectural plan" characteristic of this region (1971, p 444). Does not a "plan" suggest a planner? The maintenance of equilibrium and muscle coordination occurs in the cerebellum. Finally, there is the brain stem, which has several components that control the involuntary muscles — regulating heartbeat, digestion, breathing, etc.

Let us consider several aspects of the brain's uncanny ability. (Incidentally, human beings, unlike animals, are the only creatures who think about their brains!) The brain's memory storage capacity is incredible. It has been compared to a vast library. Carl Sagan has written:

"The information content of the human brain expressed in bits is probably comparable to the total number of connections among the neurons — about a hundred trillion, 10^{14}, bits. If written out in English, say, that information would fill some twenty million volumes, as many as in the world's largest libraries. The equivalent of twenty million books is inside the heads of every one of us. The brain is a very big place in a very small space" (1979, p 275).

It has been suggested that it would take a bookshelf 500 miles long — from San Francisco, California to Portland, Oregon — to house the information stored in man's brain. Does anyone actually believe that this kind of a library just happens? And yet this is the evolutionary assumption concerning the development of the human brain. A popular science journal employed this analogy.

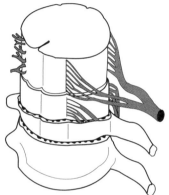

Spinal cord nerves —
Who wired this system?

"The brain is an immense computer with 110 circuits and a memory of perhaps 10^{20} bits, each of these being five to ten orders of magnitude more complex than any computer yet built. It is still more fascinating that the brain performs this work using only 20 to 25 watts compared to the six and ten kilowatts used by our large computers" (Cahill, 1981, 89[3]:105).

One writer says:

"[M]any researchers think of the brain as a computer. This comparison is inadequate. Even the most sophisticated computers that we can envision are crude compared to the almost infinite complexity and flexibility of the human brain" (Pines, 1986, p 326).

The Cray-2 supercomputer has a storage capacity about 1,000 times less than that of the human brain. One authority states that "problem solving by a human brain exceeds by far the capacity of the most powerful computers" (*Encyclopaedia Britannica*, 1989, 2:189).

No rational person subscribes to the notion that the mechanical computer just happened as the result of "happy accidents" in nature. No, the computer was obviously designed, and that demands a designer. Nobel laureate Sir John Eccles, an evolutionist, concedes the "design" evinced by the brain's amazing memory capacity.

> "We do not even begin to comprehend the functional significance of this *richly complex design* . . . . If we now persist in regarding the brain as a machine, then we must say that it is by far the most complicated machine in existence" (1958, pp 135, 136, emp. added).

If the less-complicated mechanical computer indicates design, what does this say for the infinitely more complex human brain? Richard Dawkins of Oxford University has argued that the Universe is without design. In spite of that he has written:

> "The brain with which you are understanding my words is an array of some ten million kiloneurons. Many of these billions of nerve cells have each more than a thousand 'electric wires' connecting them to other neurons. Moreover, at the molecular genetic level, every single one of more than a trillion cells in the body contains about a thousand times as much

precisely-coded digital information as my entire computer. The complexity of living organisms is matched
by the elegant efficiency of their apparent design. If
anyone doesn't agree that this amount of complex
design cries out for an explanation, I give up" (1986,
p ix).

Experiments have revealed
the brain's phenomenal memory
capacity. Under hypnosis, a brick
layer described a certain odd-
shaped brick he had laid in a building at Yale University, even though
he had laid thousands of bricks in
that structure, and the work had
been done ten years earlier (Pfeiffer, 1961, p 84).

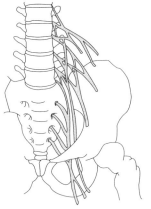

In addition to its memory
capabilities, the brain also exhib-

*Signals sent out
at 300 mph.*

its extraordinary ability in its orchestration of muscular
movements. Suppose you decide that you want to pick
up a pen and some paper from your desk. Your brain
will have to send signals to your hands, wrists, arms, and
shoulders which will direct the manipulation of 60 different joints and more than 100 muscles.

In addition to moving the muscles directionally, the
brain regulates the exact force needed for a particular
task. Opening the car door of your classic 1937 Chevro-
let requires 400 times more torque (turning force) than
dialing a rotary-style telephone. Picking up a paper clip
requires only a fraction of an ounce of force, whereas pull-
ing on your socks and shoes necessitates about 8 to 12
pounds of force. The brain compensates for multiplied

thousands of these kinds of variables in daily life.

Too, it does its work efficiently in terms of energy use. One scientist observed that "half a salted peanut provides sufficient calories for an hour of intense mental effort" (Pfeiffer, 1961, p 102). Note:

> "The average human brain weighs three pounds, consumes electrical energy at the rate of 25 watts, and occupies a volume of one-tenth of a cubic foot . . . . a machine matching the human brain in memory capacity would consume electrical energy at the rate of one billion watts — half the output of the Grand Coulee Dam — and occupy most of the space of the Empire State Building. Its cost would be in the neighborhood of $10 billion. The machine would be a prodigious artificial intelligence, but it would be only a clumsy imitation of the human brain" (Jastrow, 1981, pp 142, 143).

One of the astounding features of the brain is its ability to process and react to so many different circumstances at once. While an artist is working on a painting (using his voluntary muscles at the behest of the brain), he can: smell food cooking and know whether it is turnip greens or steak; hear a dog barking and determine if it is his dog or a neighbor's; feel a breeze upon his face and sense that rain is near; and, be reflecting on a warm friendship of the past. Even while all of this is going on, the brain is regulating millions of internal bodily activities that the person never even "thinks" about.

Logical contemplation of these facts can only lead one to agree with prominent brain surgeon, Dr. Robert White:

"I am left with no choice but to acknowledge the existence of a Superior Intellect, responsible for the design and development of the incredible brain-mind relationship — something far beyond man's capacity to understand" (1978, p 99).

Evolutionists, on the other hand, believe that bodily organs gradually evolved, by means of small changes, perhaps generated by genetic mutations, and preserved by natural selection. However, even some of these unbelievers have difficulty in dealing with this concept when it comes to something as complicated as the brain. Agnostic Robert Jastrow, a militant evolutionist, confesses: "It is not so easy to accept that theory [Darwin's theory of natural selection—WJ] as the explanation of an extraordinary organ like the brain..." (1981, p 96). Again: "Among the organs of the human body, none is more difficult than the brain to explain by evolution" (Jastrow, 1981, p 104).

Sensory Organs

In our discussion of the nervous system, it is entirely appropriate that we give some consideration to a couple of the some twenty sensory networks which play such a vital role in man's communication capacity. The fact of the matter is, we both see and hear with our brain.

(1) One of the most forceful evidences of sensory design is the eye. When Charles Darwin penned his revolutionary book, *The Origin of Species* (1859), he said more than he intended when he wrote:

> "To suppose that the eye with all its inimitable contriv-
> ances for adjusting the focus to different distances, for
> admitting different amounts of light, and for the cor-
> rection of spherical and chromatic aberration, could
> have been formed by natural selection, seems, I freely
> confess, absurd in the highest sense" (1859, p 170).

In spite of that absurdity, Darwin argued that "natural selection," with the passing of millions of years, did
produce the eye. Others have been troubled by this problem as well. Jastrow wrote:

> "The eye is a marvelous instrument, resembling a tele-
> scope of the highest quality, with a lens, an adjust-
> able focus, a variable diaphragm for controlling the
> amount of light, and optical corrections for spherical
> and chromatic aberration. *The eye appears to have been
> designed; no designer of telescopes could have done better.*
> How could this marvelous instrument have evolved
> by chance, through a succession of random events?
> (1981, pp 96-97; emp. added).

Though Dr. Jastrow argues that "the fact of evolution is not in doubt," he nonetheless confesses:

> ". . . there seems be be no direct proof that evolution
> can work these miracles . . . it is hard to accept the
> evolution of the eye as a product of chance" (1981, pp
> 101, 97, 98).

The mechanism of the eye is extremely complex.
Light images from the environment enter the eye (at
186,000 miles per second) through the iris, which opens
and shuts like the diaphragm of a camera, to let in just
the right amount of light. The images move through a
lens which focuses the "picture" (in an inverted form)

on the retina at the rear of the eyeball. The image is then picked up by some 137 million nerve endings that convey the message (at 300 miles per hour) to the brain for processing. No wonder even secular writers are prone to speak of "the *miraculous* teamwork of your eye and your brain" (Guinness, 1987, p 196, emp. added). Their vocabulary becomes rather unguarded when contemplating this phenomenon. Bioengineer, Dr. John Lenihan comments:

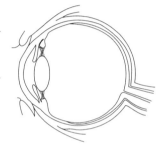

> "The eye is an exceptionally sensitive optical instrument displaying many striking features of *design* and performance; even the windscreen washers and wipers have not been forgotten" (1974, p 75, emp. added).

Lenihan is an evolutionist, so his terminology cannot be frivolously dismissed as mere creationist jargon.

The eye is frequently compared to the camera.

> "The living camera of the eye photographs fleeting images by the thousands, between one moment and the next, and it makes its own adjustments, automatically and precisely, with each change in distance, light, and angle" (Miller & Goode, 1960, p 315).

Actually, the camera was patterned after the eye, a fact admitted even by evolutionists. The Time-Life Science series volume titled, *The Body*, calls the camera "a *man-made* eye" and concedes that this optical instrument was "modeled" after the design of the eye (Nourse,

1964, p 154, emp. added). If the function of the camera demands that it was "made," does it not stand to reason that the more complex human camera, the eye, must also have had a Maker? If not, where is the fallacy in our reasoning? Note this excellent summary from Thompson:

> ". . . The eye is infinitely more complex than any man-made camera. It can handle 1.5 million simultaneous messages, and gathers 80% of all the knowledge absorbed by the brain. The retina covers less than a square inch, and contains 137 million light-sensitive receptor cells, 130 million rods (allowing the eye to see in black and white), and 7 million cones (allowing the eye to see in full color). In an average day, the eye moves about 100,000 times, using muscles that, milligram for milligram, are among the body's strongest. The body would have to walk 50 miles to exercise the leg muscles an equal amount" (Thompson & Jackson, 1990, pp 189-190).

(2) The body's hearing mechanism, the ear, is also remarkable. The ear chamber is composed of three areas: outer, middle, and inner. Sound waves enter the outer ear (at the speed of 1,087 feet per second) and pass along a tube to the middle ear. Stretched across the tube is a thin membrane, the eardrum. The sound waves vibrate this tissue. These vibrations are conveyed into the middle ear where they in turn vibrate three small bones — the hammer, anvil, and stirrup (popular names derived from the shape of these bones), that are joined together and operated by tiny muscles. The sound is thus amplified. These bones, which one authority says "are *designed* to transmit even very faint sounds," (Sedeen, 1986, p 280 emp. added), are connected to another membrane called

the oval window.

As the oval window vibrates, it generates movement within a small spiral passage, the cochlea, which is filled with liquid. The vibrations within the cochlea are picked up by 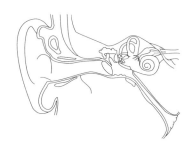 some 25,000 auditory receptors and transferred as electrical impulses, by means of the auditory nerve (with its 30,000 nerve fibers) to the brain. The brain receives these vibrations (up to 25,000 per second) and interprets them as voice, thunder, music (more than 1,500 separate musical tones), or as the thousands of other sounds that we hear daily. One authority says: "Amazingly, the inner ear, although no bigger than a hazelnut, contains as many circuits as the telephone system of a good-sized city" (Guinness, 1987, p 208).

Does anyone believe that a telephone system could design itself? The human body argues for a designer! Dr. Lenihan says that the "level of sensitivity" within the human ear is "far beyond the achievement of any microphone" and "represents the ultimate limit of performance" (1974, p 87).

The cochlea contains three tubes, called the semicircular canals, which are partially filled with fluids that move whenever the head moves. Nerve endings from these canals are connected to the brain and this, in cooperation with the muscle system, helps us keep our equilibrium or balance. The balancing ability of the auditory system has been compared to the "inertial system used in missiles and submarines" (Lenihan, 1974, p 90). So the ear-mechanism is actually designed to accomplish two functions — hearing and balance. This feature of the body demonstrates incredible planning. In the words of Dr. Lenihan: "The combination, in such a small space, of the hearing and balancing systems of the body represents a remarkable achievement of biological *engineering*" (1974, p 93, emp. added). It simply will not do to contend that "blind nature" engineered this phenomenal device.

The ear "represents a remarkable achievement of biological engineering."

— *Dr. John Lenihan*

The psalmist affirmed that God "planted the ear" and "formed the eye" (Psalm 94:9). Hearing and seeing are not developments of the evolutionary process! "The hearing ear, and the seeing eye, Jehovah has made even both of them" (Proverbs 20:12). "Our eyes and ears are transformers. They sense the light and sounds around us and turn them into electrical impulses that the brain can interpret. Each organ is *designed* to handle its own medium" (Sedeen, 1986, p 276; emp. added). Designed indeed! That's our point.

Endnotes

Asimov, Isaac (1970), *Smithsonian Institute Journal*, June.

Beck, William S. (1970), *Human Design* (New York: Harcourt, Brace, Jovanovich).

Cahill, George F. (1981), *Science Digest*, 89[3]:105.

Dawkins, Richard (1986), *The Blind Watchmaker* (New York: W.W. Norton & Co., 6th Edition).

Darwin, Charles (1859), *The Origin of the Species* (London: A. L. Burt Company).

Davies, Paul (1992), "*The Mind of God*," Omni, 14[5]:4.

Eccles, John C. (1958), *Scientific American*, September.

Encyclopaedia Britannica (1989), "Bionics," (Chicago: Encyclopaedia Britannica, Inc.).

Guinness, Alma E., Ed. (1987), *ABC's of the Human Body* (Pleasantville, NY: Reader's Digest Association).

Jastrow, Robert (1981), *The Enchanted Loom: Mind in the Universe* (New York: Simon & Schuster).

Lenihan, John (1974), *Human Engineering* (New York: John Braziller, Inc.).

Miller, Benjamin and Goode, Ruth (1960), *Man and His Body* (New York: Simon and Schuster).

Burnett, Allison L. (1961), *Natural History*, November.

Nourse, Alan E. (1964) *The Body* (New York: Time-Life).

Pfeiffer, John (1961), *The Human Brain* (New York: Harper & Brothers).

Pines, Maya (1986), in: *The Incredible Machine* (Washington, D.C., National Geographical Society).

Sagan, Carl (1979), *Broca's Brain* (New York: Random House).

Sedeen, Margaret (1986), in: *The Incredible Machine* (Washington, D.C.: National Geographic Society).

Thompson, Bert and Jackson, Wayne (1990), *Essays in Apologetics* (Montgomery, AL: Apologetics Press, Inc.), IV.

White, Robert (1978), *Reader's Digest*, September.

Chapter 5

The more we reflect on the body with which we have been blessed, the more awed we become. The body is like a mini-universe. It is integrated so wonderfully well. Each system depends upon others. None could exist without the rest. It never could have come together by "luck." Let us, therefore, continue our investigation of this phenomenal masterpiece.

The Respiratory System

The human body requires a source of power for its operational procedures. This power is obtained from food. Energy is released from food within the body by means of the metabolic process and, in order for this to be accomplished, oxygen is needed. If the body is to survive, therefore, it must have a ventilation mechanism. It does. It is called the respiratory system.

Respiration has to do with satisfying all of the body's oxygen requirements. The respiration system involves three things: (1) the intake of oxygen from the atmosphere, and the release of carbon dioxide back into the air; (2) the transportation of oxygen and carbon dioxide within the body; and, (3) the biological processes of the body's metabolic function, by which the oxygen is utilized and the carbon dioxide is released.

The respiratory system can be studied under three general categories; (1) the *air passages* to the lungs, which involve such areas as the nose, pharynx (throat), larynx (voice box), trachea (wind pipe) and bronchial tubes; (2) the *lungs*, including bronchioles (small passageways) and alveoli (air sacs); and, (3) the *skeletal and muscular apparatus* which facilitates the breathing process (e.g., the diaphragm, ribs, etc.).

This highly specialized system has been marvelously designed to support the living being. One evolutionary writer comments:

> "Lungs, heart, trachea, a bronchial tree, and connecting blood vessels all contribute to the *ingenious* breathing system that brings oxygen to the blood and removes carbon dioxide" (Schiefelbein, 1986, p 132, emp. added).

Can something be "ingenious" without there being intelligence behind it? And yet, according to materialists, this "ingenious" system is the product of unintelligent forces. This simply is not logical, and rational people will reject it.

Air moves in response to pressure differences. By means of the breathing process, air pressure in the lungs is reduced so that air is drawn from the outside atmosphere, down the passageways, into the lungs. Breathing is a very well-engineered process. Muscles attached to the ribs (the diaphragm and intercostals), together with certain other accessory muscles of the region, expand the chest by raising the ribs and lowering the diaphragm. Air rushes into the lungs, inflating them. When these muscles relax, the elasticity of the rib cage causes the chest wall to contract, and the lungs become partially deflated as the residual carbon dioxide is forced from the body. [Note:

There must be precise cooperation between the muscle
system, bone system, and
respiratory system. Too, all
of this is coordinated by
the nervous system.]

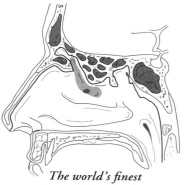

*The world's finest
air-conditioning plant!*

Every day we breathe
(15 to 18 times per minute)
in and out approximately
5,000 gallons of air. About 20% of the air is oxygen. The
respiratory system separates this oxygen from other gases
and accommodates it to the body's use.

Air enters the body mostly by means of the nose
(of course we can breathe through the mouth, but this
is not the most effective way). One writer says that "the
nose acts as an air conditioner for the respiratory system.
Every day, it treats approximately 500 cubic feet (14.2
cubic meters) of air, the amount enclosed in a small
room" (Guinness, 1987, p 218). A prominent bio-engi-
neer writes:

> "The space behind the nostrils contains the world's
> finest air-conditioning plant, combined with a detec-
> tion system of extraordinary sensitivity, which analyti-
> cal chemists are not yet able to explain, still less to
> imitate . . . . the air conditioning system of the nose
> is, in engineering terms, very well *designed*" (Lenihan,
> 1974, pp 94, 97; emp. added).

Indeed! There must have been a designer!

In the nose the rather dirty air of our atmosphere is
filtered and treated. We inhale about 20 billion particles
of foreign material daily. Nasal hair traps dust particles;

also, certain glands produce mucus (about a quart per day) which not only attracts matter, but also contains an antibacterial enzyme (lysozyme) that helps protect the body from invading germs. This waste material is moved back toward the nose and mouth by tiny, hairlike cilia that line the trachea. Eventually it is flushed from the system. Additionally, the coughing and sneezing mechanisms help rid the respiratory tract of unwanted debris.

The nasal cavities are designed to provide moisture and warmth for incoming air that may be dry and significantly below body temperature. On the way through the nasal passages, the air is preheated by a large supply of blood, "circulating like hot water in a radiator" (Miller & Goode, 1960, pp 94-95). Does a radiator suggest design?

Since air is so essential to the living machine, one would expect a skillfully engineered passageway leading to the lungs. That is precisely the situation in the trachea, a tube about $4^{1}/2$ inches long and about 1 inch in diameter. Unlike many of the passageways of the body which are composed only of muscle and other soft tissue (e.g., the digestive tract), the windpipe must remain open at all times to accommodate the flow of air. Thus, it has been designed with a series of C-shaped cartilages which provide reinforcement.

At the top of the windpipe is the larynx, or voice box. Over the larynx is an ingenious little "trap door" (the epiglottis) that prevents food from getting into the windpipe. One authority describes the cooperative action of the epiglottis and larynx as a "safety mechanism" that acts like a "valve" to prevent food from entering the windpipe during a swallow. However, if one accidentally swallows too quickly, and gets particles of food or moisture in the windpipe, another safety device, the cough, expels the invaders with a strong blast of air. The rush of air produced by a cough moves at a speed of up to 600 miles per hour (Ackerman, 1986, p 166).

Who orchestrated these movements and designed this intricate system? It is not credible to suggest that it "just happened." The mechanics of this design have to be fully operative or the individual could not survive. Evolution cannot rationally explain the swallowing/breathing process.

From the windpipe, air is drawn into the bronchial "tree," i.e., the main tubes with their multiple branches which course throughout the lungs. The air is deposited into small, spongy air sacs called alveoli. It is estimated that the lungs have 300 million or more of these air sacs.

A bronchial tree

The total lung surface exposed to air is about 45 times that of the skin surface of an adult (Beck, 1971, p 377). It is within the alveoli that the process of gas exchange, or respiration, occurs. The moist inner surfaces of the alveoli absorb oxygen molecules and diffuse them into the capillaries of the region. These oxygen molecules are then shipped to the heart, where they are pumped to the cells throughout the body. At the same time, carbon dioxide passes from the capillaries into the alveoli, and is finally exhaled. The alveoli fill and empty about 15,000 times a day during normal breathing.

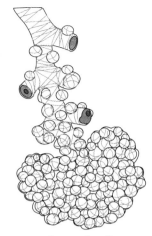

Lung alveoli

The lungs process about 6 quarts of blood every 4 minutes. [Note: The circulatory system and the respiratory system have to be fully integrated for a human being to live. These systems could not have evolved at separate times.] We don't even have to think about this exercise; it operates on "automatic" because the Creator designed it that way.

Dr. Pierre Galletti, professor of medical science at Brown University, declared: "The lung does much more than exchange gases. It may have as many as 40 different types of cells and seems to be as complex as the liver. We don't fully understand many of its functions" (see: Cauwels, 1986, p 101). Truly, in the words of the famous astronomer, Johann Kepler, we are simply attempting to "think God's thoughts after Him."

The Excretory System

The utilization of foodstuffs by a biological organism results in the production of certain waste substances which must be eliminated. Thus the human body must have an efficient system for the purging of these byproducts. The excretory system accomplishes this function in a number of ways.

The body produces several types of waste materials. Carbon dioxide, excess water, used digestive juices, worn out cells, mineral salts, nitrogen compounds, undigested food, etc., must be eliminated from the body. Earlier, we observed how the body rids itself of waste residues by means of perspiration (see: Skin System), the lower digestive tract (see: Digestive System), and the breathing process (see: Respiratory System). In this discussion, we will confine our selves to the unique design of the urinary apparatus.

The urinary mechanism consists of four major features: (1) the kidneys (the processing factories); (2) the ureters (tubes leading from the kidneys to the bladder); (3) the bladder (the receptacle for urine); and, (4) the urethra (the tube through which the urine is expelled from the bladder).

The kidneys play a vital role in the well-being of the human body. The kidneys are two dark red, bean-shaped organs located in the lower part of the back. They are about the size of an adult fist. The kidneys accomplish several important tasks: (1) they are involved in the elimination of metabolic waste products; (2) they regulate the plasma volume and the content of body water; (3)

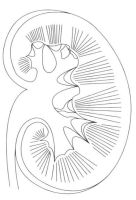

The Kidney —
"An ingenius design"

the kidneys control osmotic equilibrium, and maintain salt concentrations in body fluids; (4) they regulate the body's acid-base equilibrium; and, (5) they secrete certain hormones, e.g., renin, which is involved in the maintenance of blood pressure. Too, the kidneys produce a hormone called erythropoietin, which controls the production of red blood cells.

These functions of the urinary process work in an extraordinary fashion to preserve the life of the body. Even evolutionist William S. Beck of Harvard University was constrained to speak of the "ingenious design" characteristic of this system (1971, p 329). We must occasionally remind ourselves that even skeptics acknowledge that where there is design, there must have been a designer (Ricci, 1986, p 190).

Human kidneys consist of three layers. From the outside inward these are: the cortex, the medulla, and the pelvis. Blood flows into the cortex and medulla through the renal artery. The renal artery branches into smaller vessels which connect to a blood filtration unit called a nephron. Each nephron contains numerous small tubules (a total of about 30 miles in each kidney). Two normal kidneys contain about 2 million nephrons, which process all of the body's blood approximately every 50 minutes. Over 400 gallons of blood are pumped to the kidneys each day (Guinness, 1987, p 93). About 99% of the blood's fluid is recycled back into the body for further use. Dr. John Lenihan comments that though the kidney is a small organ, the space is "ingeniously used" since the "design" of the mechanism allows, in the tubules, a large surface for chemical transfer (1974, p 142). Yes, yes — ingenious design!

Dr. Cauwels contends that the kidneys are "greatly overdesigned" since a person can live with only one kidney, and even one organ will function with only a fraction of its capacity. This writer has a friend whose kidneys function at only about 25% of capacity. The gentleman does not feel that they are "overdesigned!"

For those who experience chronic kidney failure, several artificial kidneys have been invented. Hemodialysis usually requires several hours a week, connected to a cumbersome machine, at a cost of some $25,000 annually. God's machine is much better! Dr. Paganini, in a discussion of the design in the artificial kidney, suggested that bio-engineers need to fashion their models "so as to mimic the natural kidney," which, curiously, he calls "the *divine* prototype" (see: Cauwels, 1986, p 139; emp. added).

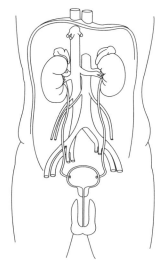

The Urinary System

The kidneys have been described as a "master chemist" whose many functions still have not been completely cataloged by physicians (Ratcliff, 1980, p 159). Lenihan, an evolutionist, confesses: "The kidney displays a mastery of chemical engineering and fluid mechanics which the technologist envies but cannot emulate. Next to the brain, it is the most complicated — and most reliable — organ in the body" (1974, p 151). The kidneys cannot be explained upon any naturalistic basis.

The two ureters, which convey the urine from the kidneys to the bladder, are tubes about 10 to 12 inches long, and about 1/4 inch in diameter. They enter the bladder at the lower portion, about 1/2 inch below the midline. The openings function like "valves" thus preventing urine from backing up into the tubes. They are constructed of three layers of tissue — fibrous, muscular, and mucous membrane. The muscular tissue contracts with peristaltic waves, which move from the upper to the lower portion of the tubes. This facilitates the flow of urine.

The bladder is a small sack (of 4 layers of tissue) which normally holds about 1 pint of urine. A sphincter muscle encircles the urethra at the bottom of the bladder. The muscle squeezes the urethra closed so that the urine is retained in the bladder. On average, about 3 pints of urine passes through the bladder daily. Significantly, urine production drops during the night (to about 1/4 of the daytime amount) so that, under normal conditions, we can get adequate rest without frequent trips to the toilet. The body's urine is a revealing index as to what is going on within the "incredible machine." The Urinalysis is one of the most valuable of all medical tests, extending back to Greek and Roman times.

The bladder has 2 sphincter valves. One opens when the bladder becomes distended as it fills with urine. Sensors signal the brain of the need to urinate when about 16 ounces of liquid have accumulated in the bladder. The other sphincter muscle, just under the former, is subject to voluntary control, which allows each person to empty the bladder at his or her convenience. Professor Beck notes that there must be precise coordination between the muscle system and the nervous system in order to accomplish this elimination process (1971, p 368). It is a remarkable example of integrated orchestration — design, if you will!

The Endocrine System

Any efficient machine must have control systems that coordinate its operations. Modern automobiles have on-board computers which function in this capacity. That idea is not new. The human body had control systems for thousands of years before man thought of the concept. The nervous system, by means of nerve impulses along its "cable network," is one type of control center; the endocrine system, with chemical stimulation, is another kind.

There is some analogy between these systems.

> "To a large extent both systems operate according to classic feedback principles: a message dispatched from a control center causes the target organ to increase or decrease its activity, and the intensity of the controlling stimulus fluctuates in accordance with incoming information on the results being achieved" (Beck, 1971, p 587).

Clearly such a function indicates intelligent planning.

The word "endocrine" derives from two Greek roots, *endo* ("within") and *krinein* ("to separate"). In contrast to glands that secrete their juices through ducts (like the sweat glands), endocrine glands are ductless. They deliver hormones (more than 100 different kinds, many perhaps yet undiscovered) directly into the blood stream, thus triggering certain bodily activities. It is clear that humans could not survive without the endocrine system. But the endocrine system could not operate without the circulatory system (Beck, 1971, p 587). The two systems must be coexistent and this presents a formidable problem for advocates of the theory of evolution. The same could be said for the relationship between the endocrine system and the nervous system. For example the adrenal medula secretes its hormones only in response to neural stimulation.

Hormones (from a Greek word meaning "to excite") have been described as "a set of coded signals" that are programmed to regulate our internal environment (Miller & Goode, 1960, p 194). A significant question would be: "Who *coded* this vital information?" The programmed specificity of the hormones is absolutely amazing, and it clearly reflects intelligent contrivance.

> "Each hormone can transmit its message only to certain target cells that have special receptors capable of recognizing that particular hormone. In a way, hormone molecules are similar to keys made so they fit some locks but not others, and receptors are like locks shaped to admit only certain keys" (Guinness, 1987, p 76).

Some of the major endocrine glands are: pituitary, pineal body, thyroid, parathyroids, thymus, adrenals, pancreas, ovaries, and testes.

The pituitary gland is about the size of a pea, and it lies in the center of the skull, just behind the bridge of the nose. The pituitary is a vital connection between the nervous system and the endocrine system. It is controlled by the brain's hypothalamus. This gland releases many hormones that affect growth, metabolism, sexual development, and reproduction. The hypothalamus/pituitary connection has been called the conductor of the "endocrine symphony" (Ratcliff, 1980, p 53). Why are men so

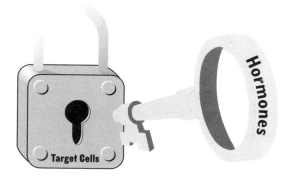

A designed system.

naive as to believe that this great symphony had no original Composer?

The pineal body is a small gland located in the brain. Evolutionists used to contend that it was merely the vestigial remains of a primitive "eye." H. G. Wells and his partners wrote: "Apparently the pineal gland is a forehead eye which first became blind and useless and then (at least in higher vertebrates) was turned to another purpose, and made into a ductless gland" (1934, p 1208). That statement is utterly ludicrous and without any "vestige" of support! It is now known that the pineal body plays a vital role in human reproduction (Bergman & Howe, 1990, pp 49-55).

The thyroid gland in the neck controls the rate of fuel use in the body, as well as body development. The parathyroids secrete a hormone that regulates the level of calcium in the blood. This is extremely important because calcium is vital for bones and teeth, the nervous system, muscle contraction, blood clotting, etc.

The thymus, located between the lungs near the top of the breastbone, was once regarded "as an evolution-

> ## The thyamus is "more complex than the defense system of any country."
>
> — *J. D. Ratcliff*

ary leftover — useless, nonproductive, a source of no good and possibly of trouble" (Ratcliff, 1980, p 64). It is now known that the thymus is the chief component of the body's defense system — "more complex than the defense system of any country" (Ratcliff, 1980, p 65).

Does anyone subscribe to the position that this nation's military defense program merely evolved by chance without any intelligent planning? One authority notes that the thymus "is responsible for the development of the immune system. In infancy, it produces cells called lymphocytes that are coded to recognize and protect the body's own tissues, while they trigger an immune response against invaders" (Guinness, 1987, p 100).

Again we must stress that a "code" is a language, and languages do not exist apart from intelligence. Clearly, Mind planned the human organism. These examples of design in man's endocrine system are clearly sufficient, to rational and honest people, to establish the existence of design in the human machine. There was a Designer! We are "fearfully and wonderfully made."

Endnotes

Ackerman, Jennifer Gorham (1986), in: *The Incredible Machine* (Washington, D.C.: National Geographic Society).

Beck, William S. (1971), *Human Design* (New York: Harcourt, Brace, Jovanovich).

Bergman, Jerry and Howe, George (1990), *"Vestigial Organs" Are Fully Functional* (Terre Haute, IN: Creation Research Society).

Cauwels, Janice (1986), *The Body Shop* (St. Louis: The C.V. Mosby Co.).

Guinness, Alma E., Ed. (1987), *ABC's of the Human Body* (Pleasantville, NY: Reader's Digest Association).

Lenihan, John (1974), *Human Engineering* (New York: John Braziller, Inc.).

Miller, Benjamin and Miller, Ruth (1960), *Man and His Body* (New York: Simon and Schuster).

Ratcliff, J. D. (1980), *I Am Joe's Body* (New York: Berkley Books).

Ricci, Paul (1986), *Fundamentals of Critical Thinking* (Lexington, MA: Ginn Press).

Schiefelbein, Susan (1986), in: *The Incredible Machine* (Washington, D.C.: National Geographic Society).

Wells, H. G., Huxley, Julian S., Wells, G. P. (1934), *The Science of Life* (New York: The Literary Club).

CHAPTER 6

How did human sexual reproduction originate? The secularist hasn't the remotest idea. Two evolutionists concede: "The origin of sex is as darkly shrouded in mystery as the origin of life itself" (Miller & Goode, 1960, p 230). But the Bible teaches that reproduction was designed by Jehovah.

God created the first humans as "male and female," and they were charged with the responsibility of being fruitful, multiplying, and populating the Earth (Genesis 1:27-28). The Creator thus provided them with the mechanism necessary for the accomplishment of this noble task.

No matter how well-designed an organism may be, if it were lacking an efficient reproductive system, it would fade into oblivion within one generation. Logic demands, therefore, that an organism's reproductive powers be fully functional. Everything must be working at once. No reproductive system could have accidentally developed in evolutionary stages.

As we have stated before, Dr. John Lenihan has been characterized as "one of the world's most eminent bio-engineers." His remarkable book, *Human Engineering — The Body Re-Examined*, is a fascinating study of the engineering techniques employed in the design of the human body. Repeatedly, Dr. Lenihan has shown that some of the most amazing engineering principles known to man were already present in man's body long before the age of modern industry. Interestingly, though, he gives but scant notice to the human reproductive system. And this was not without forethought on his part. On the final two pages of his book, the author explains:

> "Reproduction, the most important of all human activities (since without it the species would disappear), has not yet appeared in our discussion of the mortal machine. The reason for this omission is that biological replication has no mechanical analogue."

What he means is simply this. There is no machine that can illustrate the reproductive system. Man has never been able to build a machine that is able to reproduce itself. But God did! Lenihan continues:

> "Technology can prevent conception and can destroy the embryo before it reaches maturity [as is frequently done in the horrible practice of abortion - WJ] — but the engineer cannot hope to emulate the subtle, elaborate and (at least in its initial stages) uniquely enjoyable process by which the human machine replicates itself" (Lenihan, 1974, pp 206, 207).

The fascinating human reproductive system is a non-reproducible phenomenon! The male and female reproductive mechanisms are wonderfully complementary to one another. Surely this was no accident of nature; it indicates design. Miller and Goode have conceded:

> "Commonplace though it is, the fact of sexual reproduction is awesome to contemplate. And when we consider the biological mechanisms by which it operates, we find the tale almost too fantastic to believe" (1960, p 9).

The tale "too fantastic" to believe is that it could have happened by pure chance!

The Reproductive Components

Each major element of the human reproductive apparatus — both male and female — should be given consideration. The male system consists of several prime components: (1) A pair of testicles is housed in a skin pouch (the scrotum) outside of the body cavity — here sperm are produced. (2) The epididymis is a thin coiled tube (about 20 feet long) located in the scrotum beside each testicle; here the sperm develop the ability to swim. (3) The vas deferens is a duct through which the sperm pass to the urethra (a channel within the penis). (4) The penis is the organ by which the sperm are deposited from the male to the female. (5) The seminal vesicles and the prostate gland secrete substances which constitute the seminal fluid in which the sperm are carried. These fluids both protect and nourish the sperm; they also pre-cipitate contractions in the female uterus and Fallopian tubes to speed the sperm on their way toward the egg (Guinness, 1987, p 268). (6) The Cowper's glands (two pea-sized organs) produce a mucus that lubricates the urethra.

Shaped like a small egg, each male testicle contains tiny convoluted tubules (if stretched out, they would extend about 750 feet) in which the sperm are manu-factured. Sperm are produced at the rate of about 200 million every 24 hours. Special cells called Sertoli cells, which have been compared to "nurses," help to protect and nourish the developing sperm (Memmler & Wood, 1987, p 220). Each nurse cell has an elaborate "communication network" which allows it to care for about 150 sperm (Nova, 1982).

Does a "communication network" happen by chance?

Among the tubules are small cells that secrete testosterone, a male hormone which maintains the male secondary sexual characteristics, e.g., muscular development, growth of body hair, etc. The testicles (or testes) also produce some estrogen, a female hormone. "The balance between estrogen and testosterone is essential to the development of the secondary sexual characteristics" (Rayner, 1980, p 140). The production of these hormones is regulated (via the pituitary gland) by an area in the brain known as the hypothalamus.

Hence, there is an essential connection between the nervous system, the endocrine system, and the reproductive system. None could function without the others. This is clear evidence of design. One writer comments: ". . . the male reproductive system is *designed* to produce male sex cells, spermatozoa, and deposit them in the female tract . . . " (Rayner, 1980, p 142, emp. added). Yes, "designed" — and where design is apparent, there must have been a designer.

The female reproductive system has four primary features: (1) There are two small, almond-shaped ovaries about the size of walnuts which are located in the lower

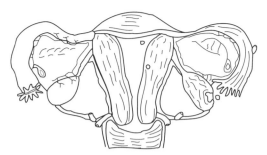

The female reproductive system.

abdomen. (2) There are two Fallopian tubes, about four to five inches long, which lead to the uterus. The interior space is about the size of a pencil lead. (3) The uterus is a muscular, hollow, pear-shaped organ a bit smaller than a woman's hand (but capable of changing size as the fetus develops) which is anchored in the lower pelvic region by ligaments. (4) The vagina, or birth canal, is a channel about three to five inches long. It is an expandable corridor by which a baby enters the outside world.

Near the vaginal opening are two small glands that secrete a lubricant, which not only facilitates sexual union, but also assists the sperm on their journey toward the ripe egg. Interestingly, during the time of ovulation, this fluid thins, thus making the "swimming" less strenuous for the sperm; at other times during the month, it thickens! Does not this circumstance suggest skillful planning?

When a baby girl is born she already has somewhere between 40,000 and 300,000 ova precursors within her ovaries. About 400 or so of these will be released during her lifetime. Ovum (egg) formation is thus a "ripening" process, rather than one of strict manufacturing. Beginning at about the age of eleven, and continuing until about the age of fifty, each month one of the female's eggs ripens. It is expelled toward one of the Fallopian tubes where it begins its descent toward the uterus.

It should be noted at this point that the ovaries also manufacture the female sex hormones, estrogen and progesterone, which are produced in cycles that are regulated by hormones from the pituitary gland in the brain. There is thus a finely orchestrated arrangement between the nervous system, the endocrine system, and the reproductive system. The mutual dependence of these systems argues for creative design and against fortuitous evolutionary development.

Conception

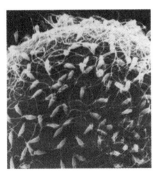

The complex orchestration that is required for the union of the male spermatozoa with the female ovum is absolutely amazing, and it could only be the result of planning. As one authority notes: "The functions of the various reproductive organs are closely coordinated" (Rayner, 1980, p 142). Coordinated? Who coordinated these functions? The Creator did.

The sperm of the male is deposited into the body of the female by means of sexual intercourse. But this is hardly the beginning of the story. As noted earlier, the sperm are manufactured in the testes.

Sperm are very tiny — about 1/600th of an inch in length, but with their vigorously lashing tail, they can swim quite effectively. A healthy male will release between 140 and 400 million sperm during each ejaculation. The spermatozoa have been ingeniously designed. Each has a "head" which houses the genetic information from the father, and a "body" which contains mitochondria that supply the energy for movement. Each has a vigorously active tail that can propel them quite rapidly in view of the fact that they are some of the smallest cells in the body. One writer compares the speed to a swimmer covering about 12 yards per second (see photo on previous page).

The Nova television feature, "The Miracle of Life," described the sperm's locomotive ability as "a highly efficient system of propulsion." Does anyone assume that the propulsion systems of our modern vehicles merely happened by chance? Certainly not. Why then should such an assumption be made with reference to this organism?

It is worthy of mention at this point that the development of sperm requires a temperature of about 95° F. Since the average body temperature is 98.6° F., there obviously must be some way to reconcile this discrepancy if the sperm are to survive. It is for this reason that the testes are located outside the body cavity within the scrotum. Amazingly, the temperature within the scrotum is controlled by the adjustment action of a smooth muscle within its walls. By means of a technical "thermostatic" process, the testes will either be drawn closer to the body, or extended further from the body, in order to maintain a proper temperature environment. This incredible design displays an astounding engineering technique.

As the sperm work their way along a rather circuitous route, they finally enter the canal of the penis, the urethra. Here they encounter what may appear to be a problem, since, as we observed in an earlier chapter, the urethra also serves as a passageway from the bladder for the elimination of urine from the body. Since urine is highly acidic, surely the sperm will be damaged!

No, for our Maker has designed a marvelous mechanism for protecting the sperm. First, the sperm are bathed in semen and, as one authority notes,

> "The fluid of the semen helps the sperm accomplish their task: It contains citric acid to dissolve the woman's cervical mucus, alkalies to buffer the vagina's acidity, sugars to give the sperm energy, and hormones to stimulate vaginal and uterine contractions that will fling some sperm far up into the woman's reproductive tract" (Schiefelbein, 1986, p 16).

Also, during the period of pre-intercourse arousal, small glands empty an alkaline fluid into the urethera, helping to neutralize any acid. Moreover, as the muscular contractions of ejaculation commence, the prostate gland shuts off the upper part of the urethera so as to prevent urine from entering the tract at this time. The Designer of this system built in a fantastic safety mechanism!

Incredibly, when the sperm are deposited, if there is an egg in the Fallopian tube, they will all "line up" and proceed in that direction; if no egg is present, the sperm will simply swim around randomly. It is almost analogous to some sort of magnetic field that has a mysterious attraction. As the sperm travel in the female reproductive tract, secretions there help dissolve their bullet-like heads, thus enhancing their ability to penetrate the ovum.

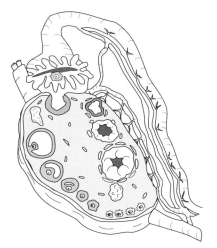

Approximately every 28 days, an egg will burst out of one the female's ovaries. The egg is about the size of the period at the end of this sentence, and is the largest human cell. The ovum is virtually tossed out into an open space in the abdominal cavity, where it is "caught" by the open end of the Fallopian tube. The open end of the tube is flared, almost like a funnel, to receive the egg. [Note: if two eggs should be released, and both are fertilized, the result would be fraternal twins.] There is obvious contrivance in this arrangement. Whereas some skeptics have criticized this mechanism as being rather haphazard, it obviously is not as evidenced by the continuing acceleration of the human population!

The Fallopian tubes are muscular organs (tubes) which lead to the uterus. They have been fashioned with finger-like projections on the interior surface. These projectiles "reach out," as it were, and gently massage the maturing little egg (called a zygote at this stage) toward the uterus. At the same time, contractions in the tube walls enhance the transportation process. Everything about this arrangement argues for design.

Normally, if the timing between intercourse and

ovulation has been integrated, somewhere in the Fallopian tube, conception will occur. When the sperm cell unites with the egg, tiny projections from the egg will grab it and pull it into the ovum's interior. The long tail (flagellum) will disconnect. At this point, the fertilized egg will form a tough covering which prevents other sperm from entering. With this union, a new human being has begun!

We might comment here that in recent years scientists have practiced what they call "in vitro fertilization." An egg is extracted from the woman and fertilized by male spermatozoa in a Petri dish (the so-called "test-tube" procedure). The fertilized egg is then transferred into the woman's uterus. Studies have found, however, that babies produced in this fashion have been associated with higher rates of miscarriage, premature delivery, low birth-weights, and various illnesses soon after birth. The "original" way is best, to say nothing of the ethical problems that are involved in some of these cases (e.g., artificial insemination by a donor – see Jackson, 1994, pp 34-36).

There is another intriguing factor that speaks eloquently of the Creator's planning. As we noted in Chapter 1, each human cell has 46 chromosomes. The chromosomes contain genes, which carry the hereditary information of the organism. Simple mathematics reveals that if the sex cells each had 46 chromosomes, the chromosomal count would be doubled with each new generation. That, of course, would never work. How is the problem solved? In a most "ingenious fashion" (Pfeiffer, 1964, p 56). While eggs are being formed in the ovaries, and sperm in the testes, a process called meiosis occurs. By this function, the ova and sperm divide and thus are left with only half of the original chromosome number, i.e., 23. Accordingly, when the spermatozoan, with its 23 chromosomes, joins with the ovum, containing its 23 chromosomes, a new person with the full complement of 46 chromosomes is formed. One has to be extremely obtuse not to see design in this brilliant plan.

We should also mention that each spermatozoan carries a sex-determination chromosome. An X chromosome has "instructions" for female traits, whereas a Y chromosome conveys the male characteristics. The egg of the female carries only X chromosomes; male sperm carry both X and Y chromosomes. Thus, if an X chromosome from the female unites with an X chromosome from the male, a baby girl results; a XY arrangement produces a boy.

As the fertilized egg makes its way down the Fallopian tube (which will take three to five days), it is dividing. Dr. Raymond Powis, who holds the Ph.D. degree in medical physiology, notes that "even at the first cell division, *dedication* and *programming* are already occurring" (1985, p 193, emp. added). Yes, but where there is a program, there must have been a Programmer. Who was it? This coding process — which has been called an "electrical language" — is directing the formation of the various features (the nervous system, legs, arms, heart, lungs,

---

Who "programmed" the "electrical language" of human development?

etc.) of this newly developing human being. At eight weeks, all the organ systems are established. An evolutionist confesses: "The phenomenon of differentiation — how cells assume their different forms — remains one of the most baffling questions of science" (Schiefelbein, 1986, p 29). They "know" it happened just by chance!

After reaching the uterus, the zygote cell cluster (of about 16 cells by now) implants itself in the uterine lining where it soon is completely covered. Inner cells become the embryo and outer cells will eventually develop into an organ known as the placenta.

The embryo (later to be called a fetus) is connected to the growing placenta by a length of tissue that will become the umbilical cord. The cord contains two arteries, and a vein, which will accommodate blood transfer between mother and child. "The architecture of the placenta lets maternal and fetal blood pass very close to each other, at times separated by no more than the thickness of a single cell, but the two blood-pools do not mix" (Powis, 1985, p 200). Who fashioned this "architecture"? The Great Architect!

The placenta is a membrane, but a very folded one, with a surface area of possibly 400 or more square feet (some have estimated 1,000 feet). It will facilitate nourishment, respiration, and waste elimination for the growing baby. As the placenta develops, it will also produce hormones, thus functioning as a "temporary endocrine gland" (Beck, 1971, p 657) that will stimulate the mother's breasts for eventual milk production.

But there is a very unusual circumstance that must be mentioned. When the embryo implants in the lining of the uterus, it has come into contact with maternal tissue, which has a completely intact immune system. This means that normally the mother's body would "fight off" the invader, and the embryo would be destroyed. It is this defense mechanism of the body that makes organ transplants so difficult.

At this point, incredible design comes into play. There are special "interfacing tissues" (called the trophoblast) that isolate the mother and the infant's systems; these are responsible for the protection of the embryo. The design is quite complex, but it is a brilliant example of the ingenious planning that has gone into the human reproductive system.

Gestation

As mentioned earlier, by eight weeks from conception the growing infant has all of its organs, but they are at various stages of development. The child must begin, however, to work as a complete system, exercising control over its structures. The infant is housed in a bag-like structure called the amniotic sac, which is filled 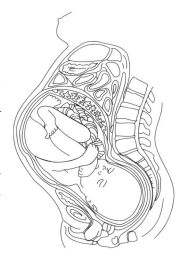 with about a quart of a clear liquid known as amniotic fluid. This fluid allows the fetus to have some movement. The baby will practice certain skills necessary for development in the outside world.

On average, the fluid is completely replaced approximately every three hours. During this continual submersion, the infant's body is protected by a layer of cheese-like material called vernix caseosa. The Creator thought of everything! Too, at this point hormones within the amniotic fluid are being activated. To use the terminology of Dr. Powis, these hormones are the "architects" of bodily change which orchestrate the formation of organs and structures (1987, p 195). Does not this suggest an Intelligence ultimately responsible for the architecture?

We must remind ourselves that during the period of gestation (normally about 280 days), the mother must supply all of the food for her child, provide all of the oxygen for her baby's developing body, and, at the same time, also eliminate the waste products from her infant. In addition, mom is having to take care of all her own needs as well. How is all of this managed? Providentially!

During gestation the mother's metabolism is altered remarkably (significantly regulated by the endocrine system). Her heart will pump more blood to supply the needs of the uterus and its little inmate. Cardiac output rises 30 to 40% above normal by the 27th week of pregnancy. Mother's lungs will work overtime to implement an increased supply of oxygen. The kidneys will take on the extra labor of excreting wastes from baby as well as mother. Increased nutritional needs for both persons will be adequately supplied, provided a good diet is consumed. The infant will take what he/she needs, even if the mother is not eating well. Can anyone honestly conclude that this all happened as a consequence of "evolutionary luck"? There are some who do.

The Birth Process

One of the mysteries of the reproductive process is the complete explanation for what triggers the beginnings of serious contractions in the uterus. What is known is the fact that near the end of the pregnancy period, the pituitary gland produces a hormone called oxytocin, which commences the stimulation of contraction in the muscles of the uterus. Too, as the time of birth nears, there is a decrease of the hormone progesterone, and an increase of estrogen; the proportionate relationship of these two hormones may be a factor in labor as well. The body seems to be so delicately balanced. It is believed that hormones from the infant may signal the maternal hormones to commence the strong birth contractions (Schiefelbein, 1986, p 49), almost as if the baby is saying: "Mom, I'm ready to be born." Once labor is started, stimuli from the cervix and vagina produce a reflex secretion of this hormone, which in turn intensifies uterine contraction.

As much as 50 pounds of pressure per square inch will be exerted. These contractions, of course, are necessary in order to push the child through the birth canal. The facts concerning how the newborn starts breathing on its own are astounding.

During pregnancy, the infant receives oxygen from the mother via the umbilical cord. The baby's lungs are collapsed and inactive. At birth, placental circulation stops and the concentration of carbon dioxide in the child's blood stimulates respiratory nerve cells in the infant's brain. The lungs expand and breathing commences. Certain blood vessels automatically close and the baby's circulatory system starts functioning as an independent unit. What if this mechanism had not worked correctly the very first time? The human race would have become extinct with Adam and Eve! Can anyone seriously believe that this precise function gradually evolved, by trial and error, over millions of years? Atheists do — though quite irrationally. Note this devastating quotation from Charles Darwin: "If it could be demonstrated that any complex organ existed, which could not possibly have been formed by numerous, successive, slight modifications, my theory would absolutely break down" (1859, p 174). It has broken down; evolutionists just refuse to face that reality.

Although there is a certain amount of bleeding in conjunction with the delivery, one authority notes that "the uterus has a remarkable ability to 'close down' its blood vessels and so prevent massive bleeding" (Rayner, 1980, p 151). Were this not the case the mother could easily bleed to death and the entire reproductive process would be for naught. Clearly, the Creator has provided numerous check-and-balance procedures to ensure the success of this divinely-designed phenomenon.

The nearest thing to instinct in humans is the sucking reflex. No baby has to be taught to nurse. The child even practices by sometimes sucking his thumb in the womb. When a baby is placed at the nipple of his mother's breast, a fascinating chain reaction is set in motion.

The sucking sensation sends nerve impulses to the hypothalamus in the mother's brain. The hypothalamus sends a message to the pituitary gland, which secretes the hormone oxytocin into the bloodstream. Certain cells within the breasts are stimulated to contract, thus forcing milk into the ducts which carry this precious liquid. Within 30 seconds after a baby begins to nurse, milk is flowing into the breast reservoirs. It is also believed that nursing accelerates the return of the uterus to its normal size. Professor Beck notes: "The uterus of a lactating mother usually becomes smaller than it was before pregnancy, whereas the uterus of a nonlactating mother often remains considerably larger" (1980, p 664).

While the practice of breast-feeding the newborn is not as popular now as it once was, still "many physicians today are convinced that there are valuable physical benefits to both mother and baby in breast-feeding" (Miller and Goode, 1960, p 245). For example, for two or three days after birth, a fluid known as colostrum is supplied by the breast. It is rich in protein and is fortified with antibodies that protect the baby against infection. It is subsequently replaced by the milk.

The Newborn

There are many aspects of human reproduction that are absolutely inharmonious with the theory of organic evolution. Not the least of these is the fact that human beings have such a prolonged physical maturation process. Whereas many animals are highly independent shortly after birth, humans are quite vulnerable. How does evolution account for this? Would not "rapid maturation" have been an excellent survival trait to retain in order to enhance the evolutionary development of the species? Of course it would.

One writer has noted:

> "Nature's animals are much more prepared for physical survival at birth, than are humans. The newborn wildebeest stands and runs off with its mother five minutes after birth. The baby elephant will follow its mother all day long within twenty-four hours after birth. The newborn human is, on the other hand, a completely helpless creature" (Cosgrove, 1987, p 98).

This is the exact reverse of what one would expect if the Darwinian scenario were true. How would infant vulnerability contribute to evolutionary survival and development? It wouldn't — if we were mere animals.

On the other hand, if human beings were designed by a loving Creator to be rational beings — nurtured by parents — so that mental, social, emotional, and spiritual traits could be developed, it makes sense that the maturing process should be delayed. Even unbelievers can hardly resist acknowledging the design in this feature of human development.

Human beings are not animals!

An evolutionist notes:

> "Psychiatrists have argued for years about the reason the human infant at birth should be so relatively inactive, compared to the young of other species that are capable of coordinated activities soon after they are born. While the answer is still not certain, it seems that this dependence of the human infant on others for his very survival over a period of years, provides him with an opportunity to undergo a long period of mental education" (Rayner, 1980, p 151).

The fact of the matter is, human beings are not mere animals. They were created in the image of God (Genesis 1:27-28); they have qualities that are unknown in the world of lower biological life.

Conclusion

The journey from conception to birth is a lengthy and complex one. It is far too complicated to have developed by random chance. The only reasonable explanation is that we have been designed by a Creator.

Endnotes

Beck, William S. (1971), *Human Design* (New York: Harcourt, Brace, Jovanovich).

Cosgrove, Mark P. (1987), *The Amazing Body Human* (Grand Rapids, MI: Baker Book House).

Darwin, Charles (1859), *The Origin of Species* (London: A. L. Burt Company, 6th edition).

Guinness, Alma E., Ed. (1987), *ABC's of the Human Body* (Pleasantville, NY: Reader's Digest Association).

Jackson, Wayne (1994), *Biblical Ethics & Modern Science*, (Stockton, CA: Courier Publications).

Lenihan, John (1974), *Human Engineering* (New York: George Braziller).

Memmler, Ruth L. and Wood, Dena L. (1987), *Structure and Function of the Human Body* (Philadelphia: J. B. Lippincott Co.).

Miller, Benjamin and Goode, Ruth (1960), *Man and His Body* (New York: Simon and Schuster).

Nova (1982), "The Miracle of Life" (a video presentation).

Pfeiffer, John (1964), *The Cell* (New York: Time Inc.).

Powis, Raymond L. (1985), *The Human Body and Why It Works* (Englewood Cliffs, NJ: Prentice-Hall, Inc.).

Rayner, Claire, Ed., (1980), *Atlas of the Body* (New York: Rand McNally and Co.).

Schiefelbein, Susan (1986), in: *The Incredible Machine* (Washington, D.C.: National Geographic Society).

CHAPTER 7

It has been argued in this book that the human body is clearly characterized by design, and therefore it must have had a Designer, namely God. Evolutionists contend otherwise. They allege that the human body is the product of a long, accidental chain of circumstances, hammered out by nature, luckily resulting in the human species. In fact, these unbelievers constantly suggest that there are many flaws in the human machine which actually argue for non-design; hence, no perfect, intelligent designer could, or would, have produced it. What shall we say in response to these charges?

First, we must admit that we do not have all of the answers to why or how the human body functions as it does. Repeatedly we have shown that even skeptical writers acknowledge the remaining mysteries of these amazing structures which we inhabit. There are secrets of cell composition, the nervous system, the reproductive system, etc., that we may never fathom. Our lack of comprehension proves nothing. One cannot argue logically from ignorance.

As we observed in a previous chapter, many organs which were once considered to be quite useless (e.g., the appendix, tonsils, etc.) are now known to be extremely valuable. But here is an example of the type of irresponsible criticism of which we speak. Richard Dawkins, in his popular book, *The Blind Watchmaker*, claims that human eyes (and those of all vertebrates) have been "wired in backwards" with the optic nerves that lead to the brain coming out of the wrong side of the retina. He says that any engineer would laugh at this design. In spite of that, he concedes that the eye works wonderfully well. He confesses: "I don't know the exact explanation for this strange state of affairs" (1986, p 93). It is just the "principle" of the thing that irritates him! If one cannot light a candle, he ought not to curse what he perceives as darkness. Elsewhere we have shown that prominent engineers, men with credentials that Dr. Dawkins might well covet, have not viewed the human eye as a laughable matter. Dawkins speaks as a God-hater, not as an expert on the human body.

Second, some features of the human makeup, which skeptics have argued are reflections of inept design, may, in fact, indicate the wisdom of the Creator and demonstrate the folly of His critics. In other words, the design may be unrecognized. Consider, for example, the fact that in the fertilization process, millions of sperm are apparently "wasted" in the venture that results in only one of them uniting with the egg. Two things must be said regarding this circumstance. First, millions of unused sperm do not logically negate the clear design that is intrinsic to the spermatozoan/egg union. Design is design, regardless of other considerations. Second, since the route from the place of sperm deposition to the location of conception is fraught with difficulty, due to other unique elements of physiological design, it is actually a manifestation of wisdom to provide for an ample sperm deposit. Even at that, fertilization is not a sure thing. Many women who desire to become pregnant do not — even under what seem to be ideal conditions.

Professor John Gerstner has an interesting analogy: "If a man were trying to catch a wild animal that was molesting his neighborhood, it would be considered a mark of intelligence for him to set a number of traps and not just one. Only one would actually be used, but to insure the accomplishment of his purpose he would be wise to set many" (1967, p 56). Let us apply this principle to the human circumstance. Though ejaculation may expel as many as 500 million sperm, 100 million of these will die almost instantly, and as few as 500 or so (or even 10) may finally reach that part of the Fallopian tube where the egg is waiting (Schiefelbein, 1986, pp 16-17). An evolutionist even admits: "The more sperm there are, the better the odds of fertilization" (Guinness, 1987, p 268). So, the employment of many sperm to finally target a single egg is not a negation of design.

Third, those who have a respect for the divine origin of the Bible acknowledge that its testimony reveals that mankind has undergone a degenerative process across the millennia of history. As a result of human rebellion, death, with all of its attendant weaknesses and ailments, has become man's abiding companion (see Romans 5:12). In viewing the human body, therefore, as thrilling as the current model is, we are not seeing this "machine" as it came from the hand of its original Maker. It has been flawed by disease, environment, and the passing of time. Adam, even in his disobedience, lived to be 930 years of age; now, however, we can only hope to reach the norm of threescore and ten, or possibly fourscore (Psalm 90:10). It was never the Lord's desire that His human creatures live forever in the sin-contaminated environs of this planet. Hence, heart trouble, cancer (or similar maladies), aging, and death will be with us as long as the human family exists. These things are not reflections upon the wisdom of God, they are commentaries on the wickedness of humanity. Elsewhere we have shown that even the innocent pay the price for this rebellion (Jackson, 1983, pp 113-120).

In this connection we must make this point. Even though there are obvious flaws in the present form of our degenerate species, that does not cancel the obvious design that is nevertheless apparent. Even atheists sometimes concede this principle. Richard Dawkins has acknowledged:

> "We may say that a living body or organ is *well designed* if it has attributes that an *intelligent* and *knowledgeable engineer* might have built into it in order to achieve some sensible *purpose*, such as flying, swimming, seeing, eating, reproducing, or more generally promoting the survival and replication of the organism's genes. It is not necessary to suppose that the design of a body is the best than an engineer could conceive of" (1986, p 21; emp. added).

It would be nice if Dawkins would accept the consequence of his own logic. Unfortunately, many are not strong on logical thinking.

Fourth, some of the criticisms of human anatomy, which infidels have advanced, are the result of stubborn stupidity, pure and simple. Woolsey Teller, one of the founders of the American Association for the Advancement of Atheism, irreverently ridiculed the body. He said: "I maintain that man makes better machines, machines

"Could somebody unzip me please?"

that work more accurately and are more dependable than those found in the human body" (*Bales-Teller Debate*, 1947, p 152).

An example of Mr. Teller's "wisdom" was evidenced by his suggestion that man would have been better designed for locomotion on all-fours, rather than upright; and fitted with wheels, rather than hands and feet. He suggested that he would have put a zipper in the human side, so that when something malfunctioned, no surgery would be necessary; one could simply unzip his body-cavity and replace the broken part. He seems not to have anticipated that such a procedure would be rather difficult to accomplish since no one would have hands with which to perform the function! How would you like a surgeon to operate on you with nothing but his two bare wheels? Teller, of course, in his assessment of the human body, parted company with the most brilliant medical experts in the history of civilization. If we may paraphrase the ancient king of Israel, "The fool hath said in his heart, there is no design in the body, hence, no designer."

The tragic fact of the matter is, some people are just determined not to believe — no matter how powerful and convincing the evidence is. Here is a typical example from infidel writer Paul Ricci:

> "Although many have difficulty understanding the *tremendous order* and *complexity of functions* of the human body (the eye, for example), there is no obvious designer" (1986, p 191; emp. added).

Observe that Professor Ricci concedes the "tremendous order" and "complexity of functions" in man's body. In spite of all that, there was no designer. You just can't do much with someone of that frame of mind. Our philosopher friend concedes that the body has "order," but he denies that it has "design." Why? Because he has admitted that where there is design, there must be a designer. He does not believe, however, that order demands an Orderer. He is merely playing a semantic game, and his quibbling cannot be taken seriously. [Note: In October of 1987, this writer debated with Paul Ricci on "the existence of God" in Chino, California.]

The Final Word

And so we must say, with calm confidence and with renewed faith, the evidence, carefully and honestly considered, leads only to the conclusion that we are creatures of God. We have been "fearfully and wonderfully made." Know that Jehovah, He is God, and it is He who has made us, and not we ourselves (see Psalms 100:3). No, we were not fashioned by nature; we are not the progeny of chance. We are His offspring, and in Him we live, move, and have our very existence (Acts 17:28-29).

As we end this presentation, we cannot do better than repeat our leading argument.

1. If it is the case that the human body is characterized by design, then it must have had a designer.

2. But it is the case that the human body is characterized by design.

3. Thus, the human body had a designer.

The human body is not an accident.
It was designed by God!

Endnotes

Bales, James D. and Teller, Woolsey (1947), *The Bales-Teller Debate* (Searcy, AR: no publisher listed).

Dawkins, Richard(1986), *The Blind Watchmaker* (New York: W. W. Norton & Co.).

Gerstner, John H. (1967), *Reasons For Faith* (Grand Rapids, MI: Baker Book House).

Guinness, Alma E., Ed. (1987) *ABC's of The Human Body* (Pleasantville, NY: Reader's Digest Association).

Jackson, Wayne (1983), *The Book of Job* (Abilene, TX: Quality Publications).

Ricci, Paul (1986), *Fundamentals of Critical Thinking* (Lexington, MA: Ginn Press).

Schiefelbein, Susan (1986), in: *The Incredible Machine* (Washington, D.C.: National Geographic Society).

Sleep – An Evidence of Divine Design

And Jehovah God formed man of the dust of the ground, and breathed into his nostrils the breath of life; and man became a living soul" (Gen. 2:7). Presently, God said: "It is not good that the man should be alone; I will make a help meet [suitable] for him" (2:18). Subsequently, ". . . Jehovah God caused a deep sleep to fall upon the man, and he slept . . ." (2:21). This is the first reference in the Bible to "sleep." Exactly what is this strange experience? Did God design it?

Science Baffled

Actually, science is puzzled about this phenomenon called "sleep." One writer has described sleep as "a nightly miracle that baffles science" (Webster, 80ff). Another comments: "On average, human beings spend a third of their lives in sleep, yet scientists do not yet know precisely what sleep accomplishes. It is presumed to serve some restorative function, but just how sleep refreshes us is unclear" (Guinness, 58). "Scientists are still seeking answers to many questions about man's need for sleep. They do not know, for example, why man cannot simply rest, as insects do. Nor have they discovered exactly how sleep restores vigor to the body" (Hartman, 418).

Theories Galore

Various opinions have been advanced to explain sleep; they all have one thing in common — they grope in the dark! As one writer observed: "There are many theories about sleep, but none is universally accepted" (Schifferes, 456). A recent article in the journal *BioScience* confessed that "modern researchers are, at the most fundamental level, as confounded by the purpose and ultimate control of sleep as were Hippocrates and Aristotle more than 2500 years ago" (Gillis, 391). Consider some of the ideas that have been advanced regarding the origin and nature of sleep.

1. Anemia — Alemaeon, a Greek physician of the 6th century B.C., argued that sleep is the result of blood draining from the head. When the cranial blood depletes to a certain level, we lose consciousness, or to express it more euphoniously, we sleep. Years ago doctors called it "cerebral anemia." It now is known that this concept is baseless.

2. Heavy Head — Aristotle, the noted philosopher of the 4th century B.C., in his work *De Somno et vigilin* (On sleeping and waking) argued that the digestion process causes "vapors" to ascend to the brain because of a higher temperature in the head. As the brain cools, these vapors descend into the heart, cooling the body's pump, producing sleep. This theory hardly needs comment.

3. Poison — The "poison" or "chemical" view alleges that sleep is the result of certain day-time waste by-products, which gradually accumulate to the point where a temporary stupor, i.e., sleep is induced. This notion is refuted by several facts. First, a person can fall asleep at any time of the day. Second, one who is sleep-ing naturally can be easily awakened — which suggests that he is not "drugged" by body poisons. Third, Sia-mese twins share the same blood system, yet one can be sleeping while the other is wide awake (Lavie, 153). This theory likewise fails.

4. Fetal Urge — Sigmund Freud, the father of "psy-choanalysis" (which is in considerable disrepute these days), contended that sleep is merely a regression from the difficulties of life. He suggested that man subcon-sciously longs to retreat to the security of "fetal life," and so he developed the sleep mechanism to accommo-date this need. One would suppose, then, that someone enlightened on this matter, as Freud obviously thought he was, could have "shucked" the sleep habit and enjoyed life awake — around the clock. He didn't!

5. Survival — Some evolutionists have argued that sleep is a development out of our animal ancestry. The claim is made that in our "pre-human" past, at night our animal kinsmen would huddle together for protection from predators. The darkness, combined with the body heat of the pack, produced a sort of trance, interrupted only by the rising sun. Over many ages this ultimately produced the crystallized habit of sleep. Thomas Edison, the great inventor of the light bulb, adopted this view and asserted: "A million years from now, we won't go to bed at all. Really, sleep is an absurdity, a bad habit" (quoted by Webster, 87). Edison charged that those who spend a lot of time sleeping are fools — which doesn't speak well of Albert Einstein, a long-sleeper (Lavie, 114). This view is downright silly. According to evolutionary chronology, true man has been upon the earth between two and three million years. Why hasn't he abandoned the sleep habit? The fact is, we still have the same sleep cycle that is evidenced in all the historical records of antiquity.

Divine Design

Earlier in this volume, we argued the case that the human anatomy is so characterized by "design," that it cannot possibly have evolved through a series of accidental circumstances. We are not a library of freak occurrences. Rather, as David, king of Israel, humbly declared: We have been "fearfully and wonderfully made" (Psa. 139:14). We do not believe there is any naturalistic explanation which accounts for the origin of sleep. We affirm that "sleep" is a mechanism, designed by God, to facilitate the well-being of certain forms of biological organisms, including man.

The Benefits of Sleep

Some still contend that sleep is non-essential for the welfare of humans. One writer recently argued that "sleep serves no important function in modern man and that, in principle at least, man is capable of living happily without it" (Meddis, Preface). The assertion is ludicrous; were it not for this "rest" provision we could not survive very long. Strange things happen when a person is deprived of sleep. After a day or two, mood changes (e.g., depression) become apparent. As days go by, experimental subjects often hallucinate and are even prone to violence. The record for staying awake is 264 hours.

Though sleep appears to have been primarily designed for the health of the brain, as we shall subsequently observe, there are numerous physical side-effects as well. Consider the following from Miller and Goode: "What happens in the body when we go to sleep, we know in considerable detail. There is a general slowing down of all the body's rhythms, a diminuendo of all its processes. Heartbeat and respiration retard to a leisurely pace; blood pressure and temperature fall to a lower level; the level of adrenaline in the blood and the volume of urine also fall" (299). Another writer notes: "Sleep restores energy to the body, particularly to the brain and nervous system" (Hartmann, 418).

Sleep also assists healing. Dr. Justus Schifferes, former Director of the Health Education Council, states that "sleep is more than a time of rest and relaxation. It is also a time of recuperation and repair, of growth and regrowth. During the normal course of living, cells of the body wear out and must be replaced. This regeneration takes place more rapidly during sleep. It has been shown, for example, that the epithelial cells of the skin divide and make new cells about twice as fast during sleep (457).

It is believed, however, that sleep performs its most powerful "magic" on the brain. This appears to be suggested by the fact that those who are deprived of sleep over several days experience minimal physical damage as compared to the mental turmoil that afflicts them. John Pfeiffer cites a study done on several hundred soldiers who stayed awake for more than four days. Medical examinations afterward revealed no significant physical debilitation. "Sleeplessness has its most important effect on one organ, the brain" (65).

Some experimental evidence suggests sleep "seems to activate the immune system" (Davis, 77). Studies have indicated that long term sleep deprivation can precipitate fatal blood infections in laboratory animals.

The brain is a paradox. It needs sleep, but it does not sleep. The fact that the brain is quite active during sleep is demonstrated in a couple of ways. First, it is the "computer" that orchestrates all of the body systems, keeping them running on "automatic," even when we are not consciously thinking about these functions. Second, "dreaming" reveals that the brain is still active during sleep. In fact, some folks have been quite creative during their sleep time. Longfellow dreamed his poem, "The Wreck of the Hesperus." Sir Isaac Newton composed some of his mathematical formulas during slumber. Professor Norbert Wiener of the Massachusetts Institute of Technology — the man who was mostly responsible for the development of the electronic calculator, frequently would jolt out of bed at night and write down the solution he had dreamed to some problem.

Modern researchers are of the opinion that sleep helps keep "the brain's nerve networks up to par." This might explain why it is so difficult to think clearly when one has been deprived of sleep. Dr. Mark Mahowald, a neurologist and a specialist in sleep disorders, says: "In a sense, sleep serves as an all-systems run-through that keeps the brain at optimal functioning" (Davis, 78).

Some suggest that sleep provides the brain with "cleanup time" in which the jumbled activities of the day are sorted and stored, much as in a computer.

Joel Benington and Craig Heller, scientists at Stanford University, are working on a theory that the brain is fueled by glucose, and when this fuel depletes to a certain level, after hours of vigorous mental activity, changes occur in a substance known as adenosine, which trigger the sleep urge. During deep sleep, the glycogen storehouse is replenished (Gillis, 1996). This theory hasn't been wholly confirmed, but its authors believe it has merit.

The emotional benefits of sleep hardly need elaboration. Perhaps they are best summed up in the words of the celebrated Shakespeare —

> Sleep that knits up the ravell'd sleeve of care,
> The death of each day's life, sore labour's bath,
> Balm of hurt minds, great nature's second course
> Chief nourisher in life's feast.

Where the Evidence Points

As one reflects upon the matters discussed above, two facts stand out clearly. (1) Sleep is an absolute necessity for human existence. (2) Man has a long way to go in understanding this phenomenon. As one writer puts it, sleep is a "complex" behavior with probably no single, simple explanation (Gillis, 393).

How can anyone, with a reasonably modicum level of rationality, argue that this beneficent and complex experience simply evolved by chance? How can a lucid person believe that: "More than three billion years ago, evolution discovered the biological clock of blue-green algae, a clock which would force us to fall asleep in a regular cycle . . ." (Lavie, ix)?

For every effect there must be an adequate cause. The data associated with sleep eloquently argue the proposition that there was an intelligent Cause for this experience. There are too many tell-tale evidences that reflect "design" in the process. And, as we have observed many times before, even skeptics concede that "everything designed has a designer" (Ricci, 190). Thank God for this provision. Sleep well!

Endnotes

Davis, Susan (1996), "Why We Must Sleep," *American Health*, April.

Gillis, A.M. (1996), "Why Sleep?," *BioScience*, June.

Guinness, Alma E., ed. (1987), *ABC's of The Human Body* (Pleasantville, NY: Reader's Digest Assoc.).

Hartmann, Ernest (1979), "Sleep," *The World Book Encyclopedia* (Chicago: World Book-Childcraft), Vol. 17.

Lavie, Peretz (1996), *The Enchanted World of Sleep* (New Haven: Yale University Press).

Meddis, Ray (1977), *The Sleep Instinct* (London: Henley &Boston).

Miller, Benjamin and Goode, Ruth (1960), *Man And His Body* (New York: Simon & Schuster).

Pfeiffer, John (1961), *The Human Brain* (New York: Harper & Brothers).

Ricci, Paul (1986), *Fundamentals of Critical Thinking* (Lexington, MA: Ginn Press).

Schiffers, Justus J. (1977), *The Family Medical Encyclopedia* (New York: Simon & Schuster).

Webster, Gary (1957), *Wonders of Man* (New York: Sheed & Ward).

Appendix II

The "Eye" of the Evolutionary Storm

Earlier in this volume, we called attention to the amazing design characteristic of the human eye. Such design eloquently argues for a designer, namely God. Further, it disputes the baseless notion that random processes in nature could produce such a phenomenal instrument.

Even staunch evolutionists have been forced to acknowledge that it is almost impossible to believe the eye could have developed by chance. Charles Darwin, who did more to popularize evolution than anyone, wrote:

> "To suppose that the eye with all its inimitable contrivances for adjusting the focus to different distances, for admitting different amounts of light, and for the correction of spherical and chromatic aberration, could have been formed by natural selection, seems, I freely confess, absurd in the highest sense" (*The Origin of Species*, London: A.L. Burt, Co.,1859, p. 170).

Dr. Robert Jastrow, an agnostic and ardent evolutionist, rephrasing Darwin, likewise felt impelled to confess:

> "The eye appears to have been designed; no designer of telescopes could have done better. How could this marvelous instrument have evolved by chance, through a succession of random events?" (*The Enchanted Loom*, New York: Simon & Schuster, 1981, pp. 96-97).

Dawkins' Charge

In chapter 7, we mentioned that evolutionists occasionally appeal to several alleged "flaws" within the body, which they claim should not be there had man been designed by an intelligent Creator, as affirmed in the Bible.

One of those we cited was Richard Dawkins, lecturer in animal behavior at Oxford University. Dawkins, an atheist, wrote a book titled, *The Blind Watchmaker* (W.W. Norton, 1986), in which he criticized the way the eye in vertebrates (including man) is "wired." He asserts that the photocells in the eyes of vertebrates are "wired backwards," a circumstance, he allows, which is laughable and offensive to "any tidy-minded engineer" (p. 93). He confesses that he doesn't "know the exact explanation for this strange state of affairs." He just knows it doesn't reflect intelligent design.

In our discussion of this matter, we observed that it is quite unwise to criticize the features of the human body. We really are just on the threshold of understanding how this amazing "machine" works. Many a critic of this biological masterpiece has ended up with a red face.

Denton's New Study

Dr. Michael Denton is the Senior Research Fellow in Human Molecular Genetics at the University of Otago, Dunedin, New Zealand. He has specialized in the genetics of human retinal disease. Though he is not a "creationist," he has, nevertheless, been vocal in his criticism of the theory of evolution (*Evolution: A Theory in Crisis*,

1984, and *Nature's Destiny*, 1998).

Dr. Denton recently authored an essay titled, "The Inverted Retina: Maladaptation or Pre-adaptation?," which was published in the Winter 1999 edition of the journal, *Origins & Design* (pp. 14-17). There, this respected scientist takes on Richard Dawkins — head-to-head. In a detailed discussion concerning the "wiring" of the vertebrate retina, Denton argues that:

> "[C]onsideration of the very high energy demands of the photoreceptor cells in the vertebrate retina suggests that rather than being a challenge to teleology [the concept of design] the curious inverted design of the vertebrate retina may in fact represent a unique solution to the problem of providing the highly active photoreceptor cells of higher vertebrates with copious quantities of oxygen and nutrients."

Denton introduces several lines of argument to buttress his case, and then says:

> "Taken together, the evidence strongly supports the notion that the inverted retina and its major consequence (the positioning of the photoreceptors in the other section of the retina where they are in intimate contact with the choriocapillaris) is a specific adaptation designed to deliver abundant quantities of oxygen to the photoreceptor cells commensurate with their high energy demands — especially in metabolically active groups such as birds and mammals. Rather than being a case of maladaptation, the inverted retina is probably an essential element in the overall design of the vertebrate visual system."

Dr. Denton continues:

"The more deeply the design of the vertebrate retina is considered the more it appears that virtually every feature is necessary and that in redesigning from first principles an eye capable of the highest possible resolution (within the constraints imposed by the wavelength of light) and of the highest possible sensitivity (capable of detecting an individual photon of light) we would end up recreating the vertebrate eye — complete with an inverted retina and a choriocapillaris separated from the photoreceptor layer by a supportive epithelium layer and so forth."

Finally, Denton concludes:

"It would seem that rather than being one of the classic 'evidences' for undirected evolution and for maladaptation [as alleged by Dawkins], the inversion of the retina is in fact highly problematic in terms of undirected models of evolution . . . . It is evidence for design and foresight in nature rather than evidence of chance."

The inspired psalmist affirmed that it was God who "formed the eye" (Psa. 94:9). Any "tidy-minded" infidel who thinks he has found a flaw in the fundamental design of the Creator's handiwork, had better use the eyes the Lord has given him, and "look again." As for the rest of us, we will give thanks unto Him who fearfully and wonderfully made us (Psa. 139:14).

Appendix III

The God Who Heals

"I am Jehovah who heals you." It was a remarkable statement. The Hebrew people had been delivered from the bondage of Egypt by the mighty hand of God. On the eastern side of the Red Sea, they wandered into the wilderness of Shur. There, they encountered the "bitter" waters of Marah. But, by a miracle, the waters were made "sweet" — a "healing," so to speak. In that connection, the Lord gave the Hebrews a "test" ordinance.

If you will diligently listen to the voice of Jehovah your God, and will do what is right in his eyes, and give ear to his commands, and keep all his statutes, I will put on you none of the diseases which I have put on the Egyptians: for I am Jehovah who heals you (Ex. 15:26).

This did not imply, of course, that no Israelite would ever become ill — even if he kept the law of God perfectly (which was not anticipated). Death is the common lot of all men (Rom. 5:12), and disease is a preliminary companion to that terminal event. It is not the will of God that humanity live in this earthly environment of sin eternally. There is a better realm where such evils do not exist (Rev. 21:4; 22:1-5).

The Lord's promise to Israel, therefore was an assurance of general well-being to those who seriously pursued his will. An old writer has noted that the Israelites, in general, enjoyed "a very good state of health" (Clarke, 378). Godliness can facilitate longevity (see Eph. 6:1-3).

In his remarkable book, *None of These Diseases*, Dr. S.I. McMillen has shown that even though Moses was instructed in all the wisdom of the Egyptians (Acts 7:22), the Hebrew health code was far in advance of that in Egypt. He concluded that the only explanation for this circumstance lies in the fact that the Pentateuch ultimately was given by God (11-24).

It goes without saying that a God who can create a man from the "dust of the ground" (Gen. 2:7), also has the ability — if he so chooses — to heal.

But there are different types of healing. There is miraculous healing wherein God has operated directly — independent of nature's laws. Then there are "ordinary" cures via the biological mechanisms of nature. In such cases the Lord has provided systems, intricately designed, to facilitate the healing of physical bodies. Let us briefly look at both of these phenomena.

Miraculous Healing

There are numerous cases, in both biblical Testaments, of God's supernatural healing power.

During the days of the prophet Elijah, a young lad was restored to life by the intervention of God's prophet (1 Kgs. 17:17-24). A similar event occurred in the days of Elisha (2 Kgs. 4:8ff). Naaman was an officer in the Syrian army, but he was afflicted with leprosy. At the Lord's bidding, however, he dipped himself seven times in the Jordan and was cured immediately (2 Kgs. 5:1ff).

There is even greater prominence given to healing in the New Testament. There are some twenty-six miracles associated with healing in the ministry of Jesus alone. The Lord could instantly cure blindness (Jn. 9:1ff), deafness (Mk. 7:31-37), hemorrhaging (Lk. 8:43-48), a withered hand (Mt. 12:9-14), leprosy (Lk. 17:11-19), or an amputated ear (Lk. 22:49-51).

In addition, there are healing miracles scattered throughout the book of Acts (2:43; 3:7; 5:1-5, 12-16, etc).

It is not the purpose of this article to develop an extensive argument for the validity of healing cases in the New Testament record. We simply note in passing: 1) The healing incidents in the ministry of Christ are entirely credible due to the vast volume of evidence that demonstrates the divine essence of these documents. 2) Even the enemies of Jesus conceded his healing power (cf. Mt. 12:22-24). 3) The supernatural cures effected by the Savior were never characterized by bizarre circumstances, as with counterfeit "healings" — in both ancient and modern times. 4) The healings performed by Jesus were never associated with monetary enrichment, as are Pentecostal cases today.

Though many have been led to believe otherwise (and in vain, we must add), miraculous healing is not being performed in this age. While claims of such are made profusely, nothing today rivals the sort of healings which the Son of God accomplished in the first century. I have attended the "healing" services of various Pentecostal groups many times over the years. I have observed scores of folks go through prayer lines with mangled bodies. Not once have I ever seen what even remotely could be characterized as a "miracle." I have seen hundreds "claim" healing when there was no observable malady. Modern healers are pathetically impotent!

Supernatural healing — which was designed to confirm revelation (Mk. 16:17ff; Heb. 2:2-4) — ceased by the end of the apostolic age, when the New Testament documents were finished (cf. 1 Cor. 13:8-10). It is vain to look for healing miracles today.

Healing By Design

Just because we deny that supernatural healings are being experienced in this age, that does not mean we do not praise our Creator for the incredible healing mechanisms that are featured in the human body. Moreover, we are confident, in the providential scheme of things, that the Lord can work through means to facilitate well-being, though, as suggested earlier, it is not his intention that we remain eternally mortal.

Just for a moment, imagine the following conversation. A wife says to her husband, "Honey, the kitchen ceiling is leaking; would you repair it?" To which he replies: "Never mind, dear; it will fix itself." Yeah, right!

What would you think of a machine that was able, time-after-time, to "repair itself." And yet that is precisely the circumstance that prevails in the human body.

A cut on your finger triggers one of the most baffling physiological processes known to modern medicine. The "healing" phenomenon, which takes place gradually in several well-defined stages, and which will even restore your fingerprint, is not understood fully even in our modern world of medical knowledge.

Physicians scratch their heads in attempting to find the words to describe it. In a book titled, *ABC's of the Human Body* (published by the Reader's Digest Association), the body's healing ability is called "a miracle" (Guinness, 99). Similarly, Miller and Goode, two evolutionary science writers, describe one phase of the healing process as "miraculous," an "astonishing performance" at which "scientists do not cease to marvel" (43).

These authors are not using the term "miracle" in the biblical sense; they are simply grasping for jargon to conceal their ignorance of how this terribly complex process came to be. The late Dr. William S. Beck of Harvard, a prominent apologist for Darwinism, described the initial stage of the healing mechanism as "among the most complex and interesting self-regulating processes in physiology." It was a phenomenon, he asserts, that "had to emerge in evolution if species were to survive" (265). The baffling question is: How did species survive while this process was developing over those eons of evolutionary time?

Moreover, are we to believe that this intricate mechanism was forged blindly by "Mother Nature," through a materialistic, evolutionary process—when the most brilliant minds of science today do not fathom how "healing" occurs? Incredible!

There are five phases in the healing process of a skin wound. Beck describes it as a "complex process requiring the coordinated functioning of many tissue elements and regulatory agencies" (748). That statement literally screams, "Design!" — a term, in fact, incorporated into the title of Professor Beck's textbook, *Human Design*. Of course, "where there is design, there must be a designer." This is fundamental logic.

The first phase of healing is the coagulation of the blood. Blood is the life stream of the body. Why does it flow easily throughout some 100,000 miles of "pipeline" within the body, only to mysteriously solidify (congeal) when a cut occurs? All of the elements for coagulation are present at any time, yet only under the most precise of circumstances is the "clotting" contrivance triggered. Who designed this "timing" device?

The "clot" is a tough, solid meshwork of fibrin strands which entrap red blood cells. This is the first line of defense as the deeper healing process begins. Subsequently, "inflammation" is generated (redness and swelling). Capillaries permeate the area, facilitating plasma and white cells which are imported to fight infection. Collagen fibrils develop. A scab forms to protect the region as the wound begins to contract internally.

As healing continues there is a tenfold rapidity in the multiplication of cells. It is as if a computer program signals — "full speed ahead!" A regenerative function is inaugurated as epidermal and connective tissue are formed. Eventually the parameters of the cut are sutured together by a tight "scar." Long after the wound appears to be "well," there still is vigorous healing activity beneath the surface.

To believe that this highly specialized operation developed "just by accident" defies the laws of critical thinking. It is the fantasy of fools.

Facilitating God's Healing Process

This writer once engaged in a debate with a representative of the "Church of the Firstborn." This cultish group repudiates the use of medicines and physicians. One member of the congregation in that community, very sincere but woefully deluded, had allowed his grievously ill baby to die, rather than consult a doctor. Curiously, the same gentleman had used a veterinarian when his milk cow became sick. [Is not a child "of more value" than an animal (cf. Mt. 10:31)?] Other groups are misguided also; "Christian Scientists" refuse medical treatment, and "Jehovah's Witnesses" prohibit blood transfusions.

Scripture does not sanction this ideology. Medicines are alluded to in the Bible on numerous occasions (cf. Pr. 3:8; 17:22; Isa. 1:5-6; 3:7; Jer. 30:13; 46:11; Ez. 47:12). In addition, Jesus suggested that those who are in health do not need a physician, but those who are ill do (Lk. 5:31). Too, Paul referred to Luke as "the beloved physician" (Col. 4:14), a descriptive that would not have been employed if the use of doctors is evil.

Since it is obvious that God has designed the human body with a capacity for healing (to some degree), we would do well to use common sense in working with nature's laws to accommodate the mending process. We can do this in several ways. 1) We can eat nutritious foods and get exercise. 2) We can try to get adequate rest (see Appendix I). 3) We can seek to avoid stress. 4) We can pursue the joys of serving God and others.

Conclusion

Finally, as we emphasized earlier, we must have a realistic view of life. Neither good health practices, physicians and medicines, or faith, prayer, and godly living are going to bestow eternal physical longevity. The main point is: Trust God and faithfully serve him — no matter what.

Endnotes

Beck, William S. (1971), *Human Design* (New York: Harcourt Brace Jovanovich).

Clarke, Adam (n.d.), *Commentary* (Nashville: Abingdon), Vol. I.

Guinness, A.E., ed. (1987), *ABC's of the Human Body* (Pleasantville, NY: Reader's Digest Assoc.).

McMillen, S.I. (1963), *None of These Diseases* (Westwood, NJ: Revell).

Miller, Benjamin and Goode, Ruth (1960), *Man And His Body* (New York: Simon & Schuster).

APPENDIX IV

SOME QUICK STATISTICS ABOUT THE HUMAN BODY

Blood - average adult, 5 quarts; 15 million platelets in one drop

Blood vessels - 60 to 100 thousand miles in body

Bones - 206 in the adult

Brain - adult, 3 pounds; 20 million volumes of information

Cells - 60 to 100 trillion; 1 million per square inch

DNA - information in one cell equals 1,000 volumes of 600 pages each

Ear - receives sound waves at 1,087 feet per second

Eye - receives light images at 186,000 miles per second

Food - human consumes 40 tons in lifetime

Foot - 24 bones

Heart -100 thousand times each day; 1,800 gallons blood per day

Hormones - More that 100 different kinds

Kidneys - process 400 gallons of blood each day

Liver - 3 pounds; 500 different functions

Lungs - 300 thousand air sacs; process 6 quarts of blood each 4 minutes

Muscles - 600 plus; 28 in face, capable of 250 thousand expressions

Nerve impulses - 350 miles per hour

Pain receptors - 3 to 4 million

Skin - average man has 20 square feet

Stomach - lining replaced every three days

Sweat glands - 2 to 5 million

Taste buds - 9 thousand

Teeth - 32 in adult human

Temperature gauges - 200 thousand

THE BIBLE & SCIENCE

BY WAYNE JACKSON

May one assert that the "spiritual" truths of the Bible are meaningful, but its "scientific" references are flawed? No, that is not consistent. The "sum" of the various parts of Sacred Writ are "truth" (Psa. 119:160 ASV). The Scriptures are scientifically credible. In fact, "science" never quite "catches up" with Scripture.

Wayne Jackson addresses whether science and the Bible agree or disagree. His style is logical, and yet, he brings complex issues down to a level that is easy for everyone to understand.

Parents will treasure this volume as they discuss with their children the day's schoolwork. College students assaulted by humanistic professors will find their faith in the Bible a comfort rather than something to be ashamed of. Read this book — then, give a copy to someone you love.

For more information regarding these and other materials that will strengthen your faith, visit Courier Publications at their web site or write to the address below.

Courier Publications
P.O. Box 55265
Stockton, CA 95205
http://www.christiancourier.com

Christian Courier

Are you interested in receiving current information on biblical studies? Do you want to have a convenient source of spiritual information available from reliable sources? If you answered "yes," then take a look at the *Christian Courier.*

The *Christian Courier* is a monthly journal of biblical studies written on the popular level. Anyone, from students to seniors, can benefit from the down-to-earth studies in this monthly journal.

The *Christian Courier* covers a variety of topics that are relevant to the serious bible student: archaeology, evolution, creation, science, biblical texts, difficult passages, theological studies, and much more.

For a free sample issue, write to *Courier Publications*, and we'll send it to you the same day.

Christian Courier on the Web

Subscribers to the *Christian Courier* have been elated to see a new edition made available over the internet. These articles are completely different, but have a lot of information packed into the web site. It's like having a vast library of biblical research available at your fingertips. And best of all, it's FREE!

Subscribe to the monthly printed edition, and point your browser to the free edition of the *Christian Courier.*

Courier Publications
P.O. Box 55265
Stockton, CA 95205
http://www.christiancourier.com

OTHER BOOKS BY
WAYNE JACKSON

The Bible & Science

The Acts of the Apostles: from Jerusalem to Rome

Biblical Studies in Light of Archaeology

The Mythology of Modern Geology

Evolution, Creation, and the Age of the Earth

Biblical Ethics & Modern Science

The Parables in Profile

Fortify Your Faith

Divorce & Remarriage Discussion

A Study Guide to Greater Bible Knowledge

Background Bible Study

The Human Body — Accident or Design?

The Book of Philippians

Select Studies in the Book of Revelation

The Book of Job

The Book of Isaiah

Debate with an Atheist — God & Ethics

Notes from the Margin of My Bible (Old & New Testament)

The Bible and Mental Health

Jeremiah & Lamentations

Treasures from the Greek New Testament

Jesus on Divorce and Remarriage

Courier Publications
P.O. Box 55265
Stockton, CA 95205
http://www.christiancourier.com